4
TYPICAL ELECTRON MICROSCOPE INVESTIGATIONS

Already published in this series:

1. *The Operation and Calibration of the Electron Microscope*
2. *Electron Diffraction in the Electron Microscope*
3. *Interpretation of Transmission Electron Micrographs*

Philips Technical Library

Monographs in Practical Electron Microscopy in Materials Science

4

TYPICAL ELECTRON MICROSCOPE INVESTIGATIONS

J. W. EDINGTON

Department of Metallurgy and Materials Science, University of Cambridge, Cambridge, England

PHILIPS

SBN 333 18640 0

PHILIPS

Printed in Great Britain by
Thomson Litho Ltd, East Kilbride, Scotland

PREFACE

This is the fourth of a series of monographs on electron microscopy aimed at users of the equipment. They are written as both texts and sources of reference emphasising the applications of electron microscopy to the characterisation of materials.

In some places the author has referred the reader to material appearing in other monographs of the series:

1. *The Operation and Calibration of the Electron Microscope.*
2. *Electron Diffraction in the Electron Microscope.*
3. *Interpretation of Transmission Electron Micrographs.*

Abbreviations and Symbols Frequently Used in this Monograph

BF	bright field (image)
DF	dark field (image)
CDF	centred dark field (image)
WB	weak beam (image)
DP	diffraction pattern
SADP	selected area diffraction pattern
SRO	short-range order
APB	antiphase domain boundary
$\boldsymbol{g}$	a vector normal to the reflecting plane
s	the deviation from the Bragg reflection position
ξ_g	the extinction distance
w	$s\xi_g$
$\boldsymbol{b}$	the Burgers vector of an undissociated dislocation
$\boldsymbol{b}_p$	the Burgers vector of a partial dislocation
$\boldsymbol{u}$	the line vector of a dislocation
FN	the upward drawn foil normal (normal to the specimen surface)
$\boldsymbol{B}$	the upward drawn incident beam direction in the specimen, that is opposite to the direction of the electron beam
ANO	anomalous absorption coefficient

CONTENTS

4. TYPICAL ELECTRON MICROSCOPE INVESTIGATIONS

4. TYPICAL ELECTRON MICROSCOPE INVESTIGATIONS

I. APPROACH TO PROBLEMS

Microstructural studies of metals and materials using transmission electron microscopy frequently involve the correlation of information present in both images and diffraction patterns. This process may be relatively easy if the interpretation of the images and diffraction patterns is simple. However, in many cases the interpretation of both images and diffraction patterns is difficult because a whole range of different finely dispersed crystal defects may be present simultaneously. Such problems frequently arise in complex commercial alloys when detailed quantitative interpretation of the microstructure can be extremely difficult. The following problems are typical.

(1) Separation and identification of different crystal defects that produce superficially similar images or diffraction effects.

(2) Identification and detailed analysis of different crystal defects responsible for a specific effect such as streaking in diffraction patterns.

(3) Collection of the maximum amount of quantitative information about all, or particular aspects of, the defect population of the crystal.

It is the purpose of the first part of this section to indicate general rules for obtaining detailed quantitative information from both simple and complex microstructures in a wide range of materials. Part II summarises some of the information in sections 2 and 3 to enable identification of crystal defects responsible for superficially similar features of microstructures and diffraction patterns. Finally parts III and IV contain a number of specific examples of detailed microstructural studies of a wide range of simple and complex materials.

4.1 General Rules for the Investigation of Microstructures

It is important to note that a series of detailed rules for investigation of complex microstructures cannot be formulated because of the diverse nature of the problems likely to be encountered. Each problem must be treated on its merits. However, a general approach would include extensive use of the following.

(1) CDF images (defined in section 1.5.2) using all of those features of the diffraction pattern, such as extra spots or streaks, to identify the features of the microstructure from which they arise, see sections 2.16–2.19.

(2) Detailed comparison of CDF and BF images which is the basis for the identification of specific crystal defects producing superficially similar (for example, fringe) images.

(3) Specimen tilting to determine the three-dimensional details of additional features of the SADP at or near a range of low-index $\boldsymbol{B}$, see section 2.10.1.3. This approach enables the type of crystal defect responsible for the fine structure of the SADP to be determined. Alternatively, by making use of CDF using precipitate reflections, particle shapes, densities, etc., may be obtained.

(4) Specimen tilting in a known sense using Kikuchi maps where detailed analysis of the sense of dislocation loops, $\boldsymbol{b}$ of dislocations, etc., is required, see section 2.10.1.1.

(5) Small amounts of specimen tilt to change the value of $\pm\boldsymbol{s}$, $\pm\boldsymbol{g}$ for two-beam conditions to determine the conditions for maximum visibility of a particular crystal defect, see sections 2.10.1, 2.10.2.

(6) Weak beam images to improve resolution when a high density of defects is present, see section 3.17.

(7) Imaging under a series of well-defined two-beam conditions, that is only one strong diffracted beam present in the diffraction pattern, see section 3.2.

II. DIFFERENTIATION BETWEEN SIMILAR TYPES OF IMAGE

The characteristic images of the defects discussed in section 3 may be used to identify crystal defects observed in thin foils. Clearly a large number of crystal defects such as incoherent large precipitates, dislocations, etc., may be readily identified on the basis of their overall morphology. However, there are three classes of image for which this is not true.

(1) Planar defects giving rise to fringe images.
(2) Small clusters giving rise to strain contrast images.
(3) Small clusters giving rise to structure factor contrast.

Within these classes different crystal defects can be responsible for the general form of the image but their detailed image characteristics described in section 3 may be used to identify the defect responsible for the image and in some cases to specify it completely crystallographically, for example extrinsic/intrinsic nature of a stacking fault.

In practice, the following information may be required to identify a particular defect.

(1) Behaviour in BF and CDF (with opposite $\boldsymbol{g}$) at $\boldsymbol{s} = 0$.
(2) Behaviour in BF for $\pm\boldsymbol{g}$.
(3) Behaviour in BF for different $\boldsymbol{g}$.
(4) Behaviour for through-focus series, that is underfocus–focus–overfocus sequences.

It is most important to realise that it is sometimes not possible to identify a crystal defect purely on the basis of its contrast behaviour. This is particularly important for small clusters, for example voids, precipitates, gas bubbles, dislocation loops. In these cases metallurgical information such as the behaviour of the defect on annealing is frequently essential to be certain of an identification.

4.2 Characterisation of Planar Defects

Planar defects produce fringe images if they lie on an inclined plane running from the top to the bottom of the foil as shown schematically in figure 3.29(a). The major characteristics of such images, based on sections 3.6, 3.7, 3.8, are summarised in table 4.1 and are enough to enable

Table 4.1 Important image characteristics for different types of planar crystal defect

Defect	Type of fringe	Spacing	Visibility	Characteristics	Quantitative information	Reference
stacking fault forms α fringe $\alpha \neq n\pi$	*displacement*, fringes parallel to intersection of defect plane with the foil surface $\alpha = 2\pi \boldsymbol{g}\cdot\boldsymbol{R}_F$: $\boldsymbol{R}_F$ is displacement vector $\alpha = \pm\frac{2\pi}{3}, \frac{4\pi}{3}$ or 0 for stacking fault	$\frac{1}{2}\xi_g$ at $s = 0$ $\xi_g/(1+w^2)^{1/2}$ at $s \neq 0$	best at $s = 0$ for both BF and DF	BF image symmetric, that is fringes at top and bottom of foil either both bright or dark. DF image asymmetric, that is opposite contrast fringes at top and bottom of foil	fringes invisible $\boldsymbol{g}\cdot\boldsymbol{R}_F = 0$. Use to determine direction of $\boldsymbol{R}_F$. Intrinsic/extrinsic nature determined by a number of techniques for $s = 0$ only. Simplest is use dark field image, transfer $\boldsymbol{g}$ to positive print, origin of $\boldsymbol{g}$ set at centre of fringes. If $\boldsymbol{g}$ points towards bright outer fringe fault is extrinsic. If dark fringe fault is intrinsic for class $B\{111\}$ $\{220\}$ $\{400\}$ reflections. For class A $\{200\}$ $\{222\}$ $\{440\}$ reflections reverse is true	section 3.6.1
planar precipitate with small misfit forms α fringe $\alpha \neq n\pi$	*displacement*, fringes parallel to the intersection of the habit plane with the foil surface $\alpha = 2\pi \boldsymbol{g}\cdot\boldsymbol{R}_n = 0$ or fraction; $\boldsymbol{R}_n$ is displacement vector normal to habit plane	$\frac{1}{2}\xi_g$ at $s = 0$	best at $s = 0$ in BF and DF	BF image symmetric, DF image asymmetric	fringes invisible $\boldsymbol{g}\cdot\boldsymbol{R}_n = 0$. Use to determine direction of $\boldsymbol{R}_n$. Precipitate smaller or larger than the matrix lattice may be determined using rules above for stacking fault. N.B. only valid if precipitate very thin and not diffracting	section 3.6.3
antiphase domain boundaries form $\alpha = \pi$ fringes	*displacement*, fringes parallel to the intersection of the habit plane with the foil surface $\alpha = 2\pi \boldsymbol{g}\cdot\boldsymbol{P}$, where $\boldsymbol{P}$ is displacement vector of APB. $\alpha = 0$ or π	$\frac{1}{2}\xi_g$, $s = 0$	best at $s = 0$ in BF and DF	BF image symmetric, DF image symmetric and exactly opposite of BF	fringes invisible $\boldsymbol{g}\cdot\boldsymbol{P} = 0$ (can be used to determine direction of $\boldsymbol{P}$)	section 3.7

Table 4.1 (*continued*)

Defect	Type of fringe	Spacing	Visibility	Characteristics	Quantitative information	Reference
δ boundary, for example between two ordered domains 1,2 with slightly different lattice parameters	δ, fringes run parallel to the intersection of the habit plane with the foil surface	$\sim \xi_g$ but depends sensitively on s	best for $s_1 = -s_2$, that is same value in both domains	BF image asymmetric, DF image symmetric (reverse of stacking fault behaviour)	fringes invisible when $\Delta \boldsymbol{g} = \boldsymbol{g}_1 - \boldsymbol{g}_2 = 0$, that is no difference in $\boldsymbol{g}$ across boundary. May be used to determine the direction of distortion	section 3.8
α–δ boundaries, for example thick precipitates	α–δ, fringes parallel to intersection of habit plane with the foil surface	depends on relative α–δ contributions	best at $s > 0$ but small	faint fringes—intermediate between α and δ boundaries	may be identified but difficult	section 3.9
thickness fringes	fringes following regions of uniform thickness, that is depth contours at edge of foil or grain boundary fringes when only one grain diffracting	ξ_g at $s = 0$ $\xi_g/(1+w^2)^{1/2}$ at $s \neq 0$	best at $s = 0$	fringes reverse contrast in BF and DF and join grain boundary fringes	used to determine foil thickness $= (n+\frac{1}{2})\xi_g$ black fringes, BF, at $s = 0$, where n = number of fringes	section 3.2.3.1
structure factor fringe at precipitates, voids, etc.	fringes parallel to the intersection of the habit plane to the foil surface. N.B. Cavity $\alpha = 2\pi s t_c$ (t_c = cavity thickness)		best at $s \neq 0$	invisible $s = 0$. Weak for thick specimens when absorption important	invisible $s = 0$, visible $s \neq 0$ may be used for identification when absorption unimportant	section 3.12
Moiré fringes at interfaces	fringes perpendicular to the operative $\Delta \boldsymbol{g} = \boldsymbol{g}_1 - \boldsymbol{g}_2$	$\lvert \boldsymbol{g}_1 - \boldsymbol{g}_2 \rvert^{-1}$	s small and positive with strong intensity in diffracted beam in each crystal	direction and spacing changes with operative reflection	Moiré fringes may be recognised by changing the operative reflection when the direction of the fringes will change	section 3.21

identification. In effect all that is required is comparison of either BF and CDF images or CDF images and SADP.

4.2.1 *Identification of the Defect Responsible for Fringe Images*

Previous sections include examples of the identification of stacking faults (section 3.6) planar precipitates exhibiting α fringes (section 3.11.3) antiphase (π) domain boundaries (section 3.7) and δ boundaries (section 3.8) using the criteria in table 4.1 combined with some of those described in the relevant sections in section 3. In this section we describe one method of determining the sense of a stacking fault and the identification of a mixed α–δ boundary.

4.2.1.1 *Stacking fault (after M. N. Thompson)*

Figure 4.1 shows a stacking fault which is identified, in this case, by using only a BF image and a SADP (van Landuyt *et al.*, 1966b). The method relies on the 'effective' streaking of the reciprocal lattice points perpendicular to the fault habit plane which produces satellite spots as described in section 2.17.1. The operative reflection in figure 4.1 was $(\bar{2}\bar{2}2)$ with $s < 0$ delineated by the position of the bright Kikuchi line inside this spot (see section 2.10.2). The inset diffraction spot is the $(22\bar{2})$ opposite reflection for which, because of the curvature of the Ewald sphere $s > 0$. The perpendicular direction of the lines joining the satellite spots c and d to the higher intensity matrix spot m indicates that they originate from the stacking faults of the type c and d respectively inclined on different $\{111\}$ planes, that is mc and md are perpendicular to the line of intersection of the fault planes and the foil surface—approximately perpendicular to the fringes. The location of the diffraction spots d and c to the right of m and the

$s > 0$ excitation error indicates that the slopes of the faults c and d are as depicted in figure 4.1(c). In the micrograph the top and bottom of the faults are represented by T and B.

In relation to the single fault c the following information has been established.

(1) $\boldsymbol{g}$ is $[\bar{2}\bar{2}2]$ class A, table 4.2.

(2) The relative orientation of the fault and $\boldsymbol{g}$ is as shown in figure 4.1(c).

(3) The extreme dark fringes in the positive BF print, figure 4.1(a), are dark–dark.

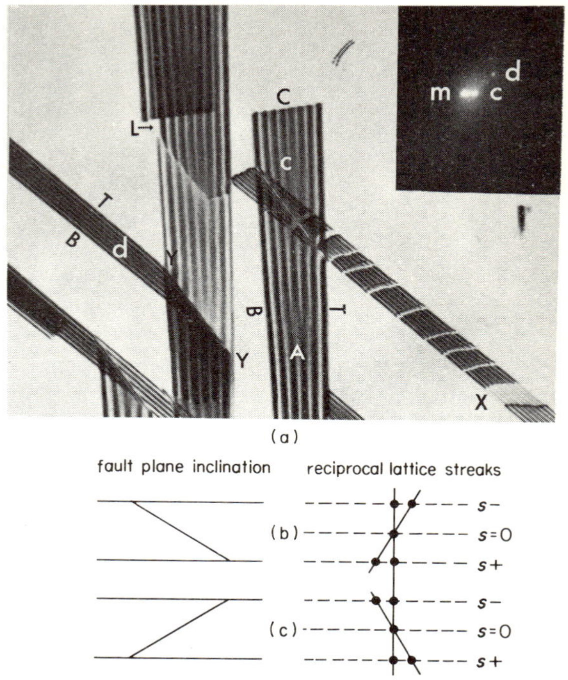

Figure 4.1 (a) The relationship between the fine structure of the diffraction spot and the BF image of intersecting stacking faults. Inset $(22\bar{2})$ diffraction spot s positive. (b), (c) The relationship between fault plane inclination, reciprocal lattice streaking, and the dependence of matrix and satellite separation on the deviation parameter s. The dashed line represents the Ewald sphere

Table 4.2

Diffraction vector	Class
[200]	A
[222]	A
[440]	A
[400]	B
[111]	B
[220]	B

Reference to tables 4.2 and 4.3, taken from van Landuyt *et al.* (1966b), establish fault c as intrinsic. Analysis of the other stacking faults is prevented by the overlapping effects which are responsible for the fringe shifts, etc., see section 3.6.1.2.

Table 4.3

$\boldsymbol{g}$ →	Bright field A		Bright field B		Dark field A		Dark field B	
Extrinsic (top)	D	D	B	B	B	D	D	B
Extrinsic (bottom)	B	B	D	D	B	D	D	B
Intrinsic (top)	B	B	D	D	D	B	B	D
Intrinsic (bottom)	D	D	B	B	D	B	B	D

——— fringe corresponding to top surface; — — — fringe corresponding to bottom surface. B and D mean bright and dark, respectively, that is the nature of the extreme fringes.
From van Landuyt *et al.* (1966b).

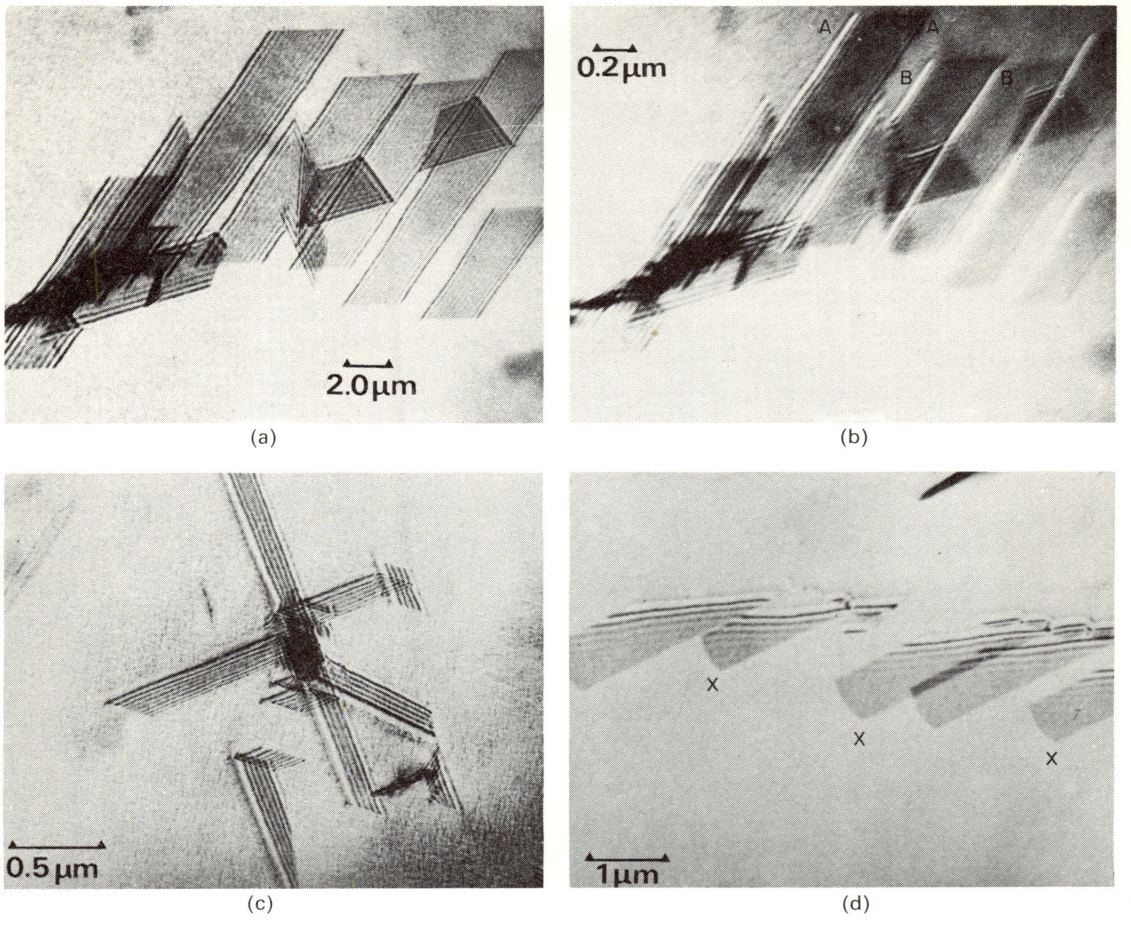

Figure 4.2 Images of plate-shaped precipitates in niobium: (a) BF; (b) CDF; (c) intersection of fringe patterns; (d) fringe systems ending inside the foil at X where features are observed in disagreement with the stacking fault hypothesis [From van Landuyt *et al.* (1966a), courtesy of *Physica Status Solidi*]

4.2.1.2 *Mixed α–δ fringes (after J. van Landuyt)*

Figure 4.2 shows the identification of mixed α–δ fringes due to van Landuyt *et al.* (1966a). Close examination of the fringes in BF and CDF reveals that their central parts are different from the outer parts, see figure 4.2(a), (c). This may be interpreted in terms of a central region of overlap of two closely spaced interfaces whereas the outer parts on both sides are associated with single interfaces, that is the geometry in figure 3.37(a). Further observations are as follows.

(1) There is no difference in background on either side of the fringe patterns.

(2) The BF image is symmetrical but the DF is asymmetrical in region AA but symmetrical in region BB, figure 4.2(b).

(3) Partial dislocations were not necessarily present along the border of the fringe patterns, figure 4.2(d).

The interpretation is as follows.

(1) The fringes are caused by thin precipitate plates, that is region II, figure 3.37(a), would be a precipitate. This is consistent with observations that these defects protrude from the foil edge.

(2) The fringes are due to overlapping α–δ interfaces. The δ component could be due to a slight misorientation of the reflecting planes in both phases, together with possible differences in ξ_g and absorption length. The α component may arise from the different phase angles produced in the different crystals.

The defects are not displacement fringes because, if the contrast were due to a shift between the crystal parts on either side of the precipitate, the region of overlap would exhibit the α fringes described in section 3.6.1, whereas the non-overlap parts would exhibit wedge fringes. Although region AA has α fringes BB does not, that is BF and DF images are the same.

The interfaces are not δ fringes because, if they were, they should have equal and opposite components as the background intensity is the same on either side. Closely overlapping δ interfaces produce an α fringe pattern in the overlapping part but the outer fringes of the images in BF would be opposite in sign, see section 3.8.1.2, which they are not.

The interfaces are not α interfaces because (1) the overlapping region should have α fringes as observed, but (2) the outer fringes should have opposite contrast in either BF if the signs of α are opposite or DF if α_1 and α_2 are both positive. This is not observed.

The interfaces do not arise from a shift in the two parts of crystal 1, figure 3.37(a), because, if the precipitate does not diffract, the region of overlap would exhibit α fringes whereas the non-overlap parts would exhibit wedge fringes, see van Landuyt *et al.* (1965). Although region AA exhibits α fringes BB does not and the BF and DF images are the same.

4.3 Characterisation of the Strain Field Images of Precipitates, Loops, Gas Bubbles, End-on Dislocations and Etch Pits

These defects produce either small black/white lobe contrast or large black/black lobes with occasional examples of large black/white lobes, see section 3.11.1.2. The characteristics of these images are discussed below and summarised in tables 4.4 and 4.5. It is important to realise that complete identification based on their contrast behaviour in BF and CDF images can be difficult. Additional metallurgical information is frequently required, for example the behaviour of the defect on annealing.

4.3.1 *Small Black-dot or Black/White Lobe Images*

The size of these images is usually $\ll \xi_g$ and they

Table 4.4 Image characteristics of crystal defects giving dot or small lobe (black/white) images $\lesssim 0.25\xi_g$ in diameter

Defect	Type of image	Visibility, BF	Characteristics, BF	Quantitative information	Reference
Frank loop <100 Å *diameter*, for example $\boldsymbol{b} = \frac{1}{3}\langle 111\rangle$ f.c.c.	lobes of black/white contrast near top and bottom of foil. At critical depths black-dot contrast	best at small values $s > 0$ slightly dark background, that is thickness $= (n \pm \frac{1}{4})\xi_g$, near dark thickness fringes	black/white contrast reverses depending on depth, that is at $0.3\xi_g$ $0.75\xi_g$ and does not occur below $1.25\xi_g$. Black dots at $0.3\xi_g$ and $0.75\xi_g$ and within centre of foil independent of $\pm\boldsymbol{g}$. The line $\boldsymbol{l}$ joining centre of black and white lobes is parallel to projection of $\boldsymbol{b}$	with *considerable* care the sense and Burgers vector of loops may be determined	section 3.5.5
small loops with a shear component	images complex and interpretation is not covered by theory at present				
small precipitates $\boldsymbol{g}\varepsilon r_0^3/\xi_g^2 < 0.2$ (a) spherically symmetric strain fields	small black/white lobes	as for Frank loops	black/white contrast changes with depth as for Frank loops (see above) and the line $\boldsymbol{l}$ joining centre of the black and white lobes is parallel to $\boldsymbol{g}$	with *considerable* care the sense of the strain field may be determined	section 3.11.1
(b) asymmetric strain fields, that is non-spherical particle and/or elastically anisotropic matrix	as for (a)	as for (a)	as for (a) but $\boldsymbol{l}$ not always parallel to $\boldsymbol{g}$	with *considerable* care it should be possible to determine the sense of strain field using the rules for (a). However, no theoretical background	section 3.11.1
gas bubbles with spherically symmetric or asymmetric strain fields	as for (a)	as for (a)	images have same form as (a), (b) above. In the case of bubbles $d \lesssim 0.3\xi_g$ these may only be visible under dynamical contrast conditions. However, show characteristic through-focus properties, that is overfocus properties dark spot, underfocus bright spot	with considerable care it should be possible to determine the sense of the strain field as for (a), (b) above	section 3.13

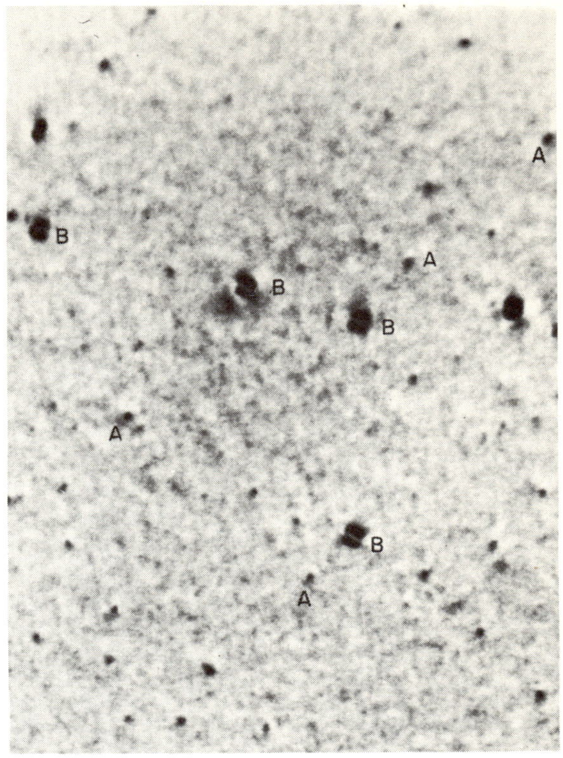

(a)

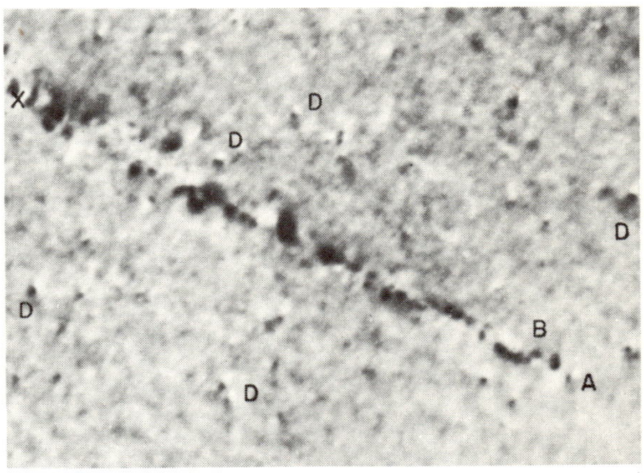

(b)

Figure 4.3 BF electron micrographs of TiC particles B. (a) In quenched and aged Ni–4.47 wt % Ti–1.96 wt % C alloy in the presence of small Frank loops A. (b) A row of very small particles intersecting the bottom surface of the specimen at A

can arise from precipitates, gas bubbles, dislocation loops or etch pits with a strain field, see table 4.4. The image characteristics of etch pits with strain field have not been studied and cannot be included in table 4.4. Consequently, the best method of detecting such a defect would be stereo microscopy which should demonstrate that the defects lie at the surface. The important image characteristics of the remaining defects are listed in table 4.4 and they may be supplemented by details from sections 3.5.5, 3.11.1 and 3.13.

It is important to realise that, from an experimental and theoretical viewpoint, the contrast behaviour of the following defects is reasonably well understood: Frank loops for isotropic elasticity; precipitates and gas bubbles with spherically symmetrical strain fields. For the other defects complete information is not yet available.

The following information may be readily obtained.

(1) Frank loops may be distinguished from precipitates/bubbles with spherically symmetrical strain fields because $\boldsymbol{l}$ is always approximately parallel to the projection of $\boldsymbol{b}$ with different $\boldsymbol{g}$ whereas $\boldsymbol{l}$ is parallel to $\boldsymbol{g}$ in the latter case.

(2) The sense of the displacement associated with Frank loops, precipitates and bubbles with spherically symmetrical strain fields may be determined using the approach outlined in section 3.5.6.2.

The following experiments are extremely difficult.

(1) To distinguish between gas bubbles and small precipitates with spherically symmetrical strain fields. In this case the specimen may be tilted to kinematical conditions and a through-focus series obtained which may enable the bubbles to be identified. Alternatively, extra spots may be present in the SADP from precipitates if the volume fraction is large.

(2) To distinguish between precipitates with edge loop type asymmetric strain fields, for example precipitate plates and Frank loops.

(3) Any problems associated with small plate defects (loops, precipitates) with a shear component of the strain field.

4.3.1.1 *Small precipitates versus Frank loops (after R. B. Scarlin)*

A very good example of this is shown below describing the differentiation between Frank loops and fine misfitting 'matrix dot' TiC precipitates in a nickel-base alloy. In a number of materials, MC carbides such as TiC form small, finely dispersed, cuboidal precipitates (edge length $\lesssim 100$ Å) with a cube–cube orientation relationship to the matrix. Because of the misfit strain, these defects exhibit small black/white strain field images. Figure 4.3(a) shows a region of a Ni–2 wt % Ti–0.3 wt % C alloy aged at 600 °C for 200 hours. Two types of defect are present, large black/black lobes of contrast at B and small black/white contrast lobes at A. There are no extra spots or streaks in SADP. The black/black lobes are typical of the strain field image from a precipitate with $P_s \gtrsim 0.2$, section 3.11.1.2, whereas the small black/white images could be either fine-scale precipitate or Frank loops. In fact they were shown to be $\frac{1}{3}\langle 111 \rangle$ vacancy Frank loops, see section 3.5.6.1.

Table 4.5 Image characteristics of crystal defects giving large ($r > 0.25\xi_g$) black/black or black/white lobe contrast

Defect	Type of image	Visibility, BF	Characteristics, BF	Quantitative information	Reference
precipitate with (a) spherically symmetrical strain field, $P_s = \boldsymbol{g}\varepsilon r_0^3/\xi_g^2 \gtrsim 0.2$	black/black lobes in foil, black/white lobes if precipitate lies within ξ_g of surface	best at $\boldsymbol{s} > 0$ and at thicknesses with $(n\pm\frac{1}{4})\xi_g$ which exhibit slight background contrast	image diameter $\sim\xi_g$. Large black/black lobes line of no contrast always perpendicular to $\boldsymbol{g}$. Black/white reverses on reversing $\boldsymbol{g}$ but black/black images unaffected	with care the sense of the strain field may be determined from black/white images and its value estimated from black/black images	section 3.11.2.1
(b) asymmetric strain field, that is non-spherical particle or elastically anisotropic matrix, $P_s = \boldsymbol{g}\varepsilon r_0^3/\xi_g^2 \gtrsim 0.2$	not quantified but assumed to be similar to (a)	as for (a)	the same general behaviour as (a) would be expected and has been proved for Co particles in Cu[a]. N.B. Line of no contrast not oeroendicular to $\boldsymbol{g}$ for all $\boldsymbol{g}$ and sometimes curved[b]	possible with care to determine the sense of the strain field	section 3.11.1
gas bubbles	as for (a)	as for (a)	some general features as above (a), (b) for precipitates except that the through-focus properties are underfocus bright spot, overfocus dark spot	with considerable care the sense of the strain field can be determined and the magnitude estimated using the Ashby–Brown analysis for spherically symmetrical strain fields	section 3.13
end-on dislocations	black/white lobes	best at $\boldsymbol{s} > 0$	image diameter $\sim\xi_g$, stronger images for screw than edge dislocations	the sense of the dislocation RH or LH screw may be determined	section 3.4.8
medium-sized perfect loops, 100–500 Å diameter, for example · $\boldsymbol{b} = \frac{1}{2}\langle 110\rangle$ f.c.c. $\frac{1}{2}\langle 111\rangle$ b.c.c.	resolvable loop	$\boldsymbol{s} > 0$ but small use thicknesses $(n\pm\frac{1}{4})$ ξ_g to give some background contrast	contrast is not significantly dependent on the depth of the loops in the foil. However, dot changes to double black lobe $\pm\boldsymbol{g}$	with considerable care the sense and Burgers vector of the loop may be determined. However, consistent results may only be obtained for a fraction of the loops	sections 3.5.3, 3.5.4
large perfect edge loops, diameter > 500 Å $\boldsymbol{b} = \frac{1}{2}\langle 110\rangle$ f.c.c. $\frac{1}{2}\langle 111\rangle$ b.c.c.	resolvable loop	best at small values of $\boldsymbol{s} > 0$	contrast not significantly dependent on the depth of the loop in the foil but loop image changes size $\pm\boldsymbol{g}$	the Burgers vector and sense of the loop may be obtained reasonably easily	sections 3.5.1, 3.5.2
medium-sized Frank loops $\boldsymbol{b} = \frac{1}{3}\langle 111\rangle$ f.c.c.	resolvable loop	best at small values of $\boldsymbol{s} > 0$	contrast not significantly dependent on the depth of the loop in the foil but fringes from stacking fault are not observed	the Burgers vector of loop may be obtained using $\boldsymbol{g}\cdot\boldsymbol{b} = 0$ and $\boldsymbol{g}\cdot\boldsymbol{b} = 1$ conditions but avoiding $\boldsymbol{g}\cdot\boldsymbol{b} = \frac{1}{3}\pm\frac{2}{3}$. Loop may be separated from perfect loop by reversing $\boldsymbol{g}\pm[200]$ when Frank loop reverses contrast	sections 3.5.3, 3.5.4
large Frank loops, diameter $\gtrsim$ 500 Å $\boldsymbol{b} = \frac{1}{3}\langle 111\rangle$ f.c.c.	resolvable loop containing stacking fault fringes	best at small values of $\boldsymbol{s} > 0$	contrast of fringes better if loop near top and bottom of foil and also better for first-order reflections	the Burgers vector of loop may be obtained using $\boldsymbol{g}\cdot\boldsymbol{b} = 0$ and 1 conditions but avoiding $\boldsymbol{g}\cdot\boldsymbol{b} = \pm\frac{1}{3}, \frac{2}{3}$. Sense may be determined using $\boldsymbol{g}\cdot\boldsymbol{b} = \pm\frac{4}{3}$	sections 3.5.1, 3.5.2

[a] McIntyre and Brown (1966).
[b] Degischer (1972).

These loops are concluded to arise from ion damage in the microscope for the following reasons.

(1) They lie near the surface facing the electron gun.
(2) The defects form in the alloy independently of its thermal history and they are also observed in pure nickel.
(3) They are not confined to the area irradiated by the electron beam.
(4) The loop diameter (~ 80 Å) is consistent with the formation of each loop from an individual ion impact. An 100 kV oxygen ion produced 10^3 Frenkel pairs and the maximum loop diameter produced by clustering of the vacancies, if all the interstitials escape to the foil surface, is ~ 80 Å.
(5) The loop density after 3 hours' observation in the microscope is $\sim 10^3\ \mu m^{-2}$, which gives a rate of defect production similar to that reported by Pashley and Presland (1961) for gold specimens. Assuming that each loop corresponds to an individual ion impact, the above loop production rate corresponds to an ion irradiation rate of $\sim 10^7\ cm^{-2}\ s^{-1}$, in good agreement with the measurements reported by Howie *et al.* (1966).

It is also possible to differentiate between ion damage of this type and fine precipitates with $P_s \lesssim 0.2$ which also produces black/white lobe contrast. In this particular case, because the ion damage is almost all within layer I, figure 3.27(a), whereas the precipitates are uniformly distributed throughout the foil, $\boldsymbol{g} \cdot \boldsymbol{l}$ has constant sign for loops but varies positive and negative for the precipitates. Figure 4.3(b) shows this effect clearly. Two rows of small TiC particles have been precipitated in the wake of a climbing perfect dislocation, in the manner suggested by Silcock and Denham (1969). The row of particles end at the bottom surface of the foil at A and each particle along the line AX lies at successively higher levels, for example the particles B lie within layer II, figure 3.27(a), and the sign of $\boldsymbol{g} \cdot \boldsymbol{l}$ has changed. Using the rules in section 3.11.1.1, $\boldsymbol{g} \cdot \boldsymbol{l}$ is negative level I and positive level II, figure 3.27(a), and the defects are interstitial. A significant amount of ion damage exhibiting the contrast of vacancy loops is also present in this area, for example D.

4.3.2 *Large Black/Black or Black/White Lobe Images*

The size of these images is usually $\sim \xi_g$ and they can arise from precipitates, gas bubbles, dislocation loops, end-on dislocations, stacking fault tetrahedra (table 4.5). The cases of spherically symmetrical strain fields of precipitates or bubbles, together with the images of end-on dislocations, dislocation loops, are well understood. However,

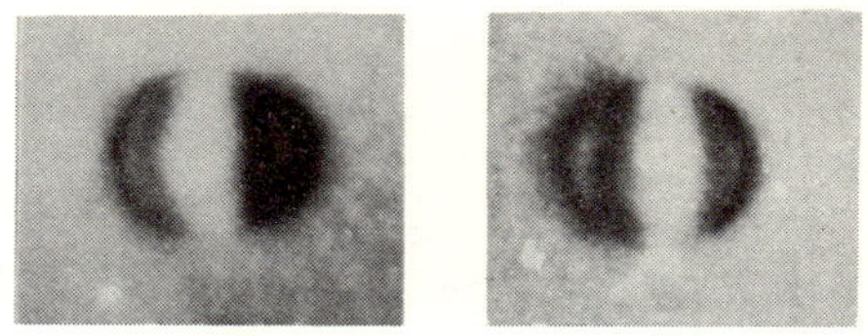

(a) (b)

Figure 4.4 The strain field contrast behaviour of a δ' Al_3Li precipitate in an aged Al–10.7 at % Li alloy for $\pm \boldsymbol{g}$ [331]

asymmetric strain fields from precipitates or bubbles have only received limited attention so specific conclusions may not be drawn except in a few cases.

The following distinctions may be readily made between the following.

(1) Dislocation loops and precipitates with strain field. The image of the loop will change size on reversing $\boldsymbol{g}$ but that of a precipitate will not. Compare figure 4.4(a),(b) with figure 4.8(d),(e). Note that only precipitates show black/white lobe contrast when near the surface because a loop usually has a smaller strain field, section 3.5.5.1.
(2) Large precipitates with black/black lobes and end-on dislocations because the latter only have black/white lobe images. However, this distinction may be avoided by using thicker regions of the foil where the dislocation end effect is less obvious and they may be detected by tilting the foil 30–40°, that is they are no longer end on. If for some reason very thin regions cannot be avoided, use a number of different $\boldsymbol{g}$ so that the loops or planar precipitate images (which are relatively unaffected by $\boldsymbol{g}$) can be identified. The end-on dislocation images should change significantly under different $\boldsymbol{g}$, for example for a screw dislocation the black/white lobes always on either side of $\boldsymbol{g}$, $\boldsymbol{l}$ perpendicular to $\boldsymbol{g}$, see section 3.4.8.
(3) Precipitates and gas bubbles. The through-focus properties of gas bubbles may be used to characterise these defects, see section 3.13.2.
(4) Large Frank and perfect loops because the former contain stacking fault fringes.
(5) Small Frank loops and perfect loops may be distinguished by obtaining images in $\pm\boldsymbol{g}$ for two $\{020\}$ reflections such that $\boldsymbol{g} \cdot \boldsymbol{b} = \pm\frac{2}{3}$ for all four $\boldsymbol{b} = \frac{1}{3}\langle 111 \rangle$. Frank loops show reversal of contrast whereas prismatic loops do not, see figure 4.5(a),(b) (Loretto *et al.*, 1966).
(6) The spherical symmetry, or otherwise, of the precipitate strain field may be detected by observing the direction of the line of no contrast for different reflections which should always be perpendicular to $\boldsymbol{g}$. Figure 4.6 shows an example of TiC particles in the nickel-base alloy discussed in section 4.3.1.1. Note that $\boldsymbol{g}$ is not perpendicular to the line of no contrast, that is the strain field

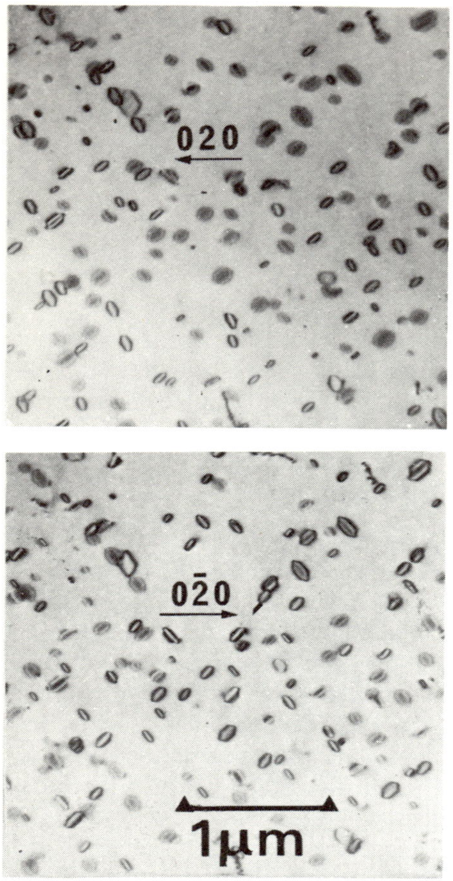

Figure 4.5 BF images of quenched-in dislocation loops in aluminium $\boldsymbol{B}$[001]. The deviation s from the Bragg position is large and the distinction between loops in good and bad contrast is obvious [From Loretto *et al.* (1966), courtesy of *The Philosophical Magazine*]

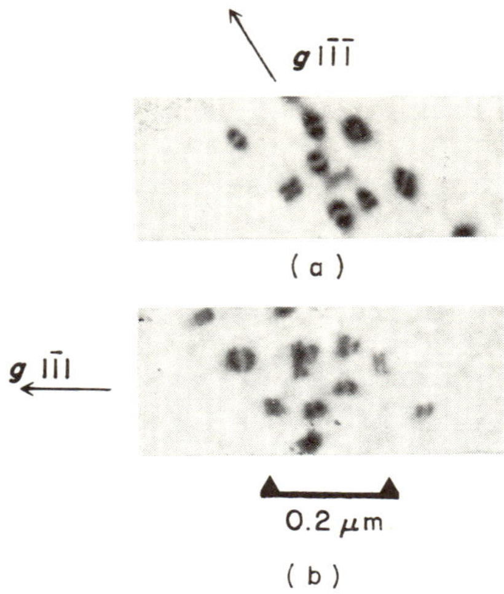

Figure 4.6 BF images of TiC particles in quenched and aged Ni–4.47 wt % Ti–1.96 wt % C observed under different reflections

is asymmetric, and the line of no contrast changes direction depending on $\boldsymbol{g}$.

(7) Note that precipitates may be most easily detected and identified from extra spots in SADP if the volume fraction is high enough. Alternatively they may occasionally protrude from the edge of the foil.

However, the following distinctions are more difficult.

(1) Between black/white precipitate images and particles near the surface and end-on dislocations. This situation is best avoided by using thicker regions and tilting through large angles to reveal the dislocation line.

(2) Any work involving non-spherically symmetrical strain fields.

4.3.2.1 *Identification of coherent precipitates and dislocation loops in a high-phosphorus stainless steel (after B. L. Eyre)*

An example of a contrast analysis including some of the defects discussed in this section is given below for a steel containing 18.18Cr–10.25Ni–0.012C–0.31P–0.60Mn–0.31Si–0.009N in wt % (Rowcliffe and Eyre, 1971). The heat treatments which produce the two defects of interest are given in table 4.6. In the following discussion the importance of metallurgical information, such as annealing behaviour, in addition to contrast behaviour, is clearly demonstrated.

Table 4.6 Heat treatments

Solution treatment	Ageing treatment	Defect observed
1100 °C, ½ hour	650 °C, 1 hour	type 1
1100 °C, ½ hour	500 °C, 1 hour	type 2
1100 °C, ½ hour	500 °C, 10 hours	type 2

Micrographs were obtained under strong two-beam conditions, see figure 4.7, and the characteristic images of these defects are discussed below.

Defects 1. These defects occur in a high density, see figure 4.7(a), and from their black/black lobe contrast could be either precipitates with strain fields, gas bubbles or dislocation loops. However, their contrast behaviour under different reflections enables some of these to be rejected, for example the following possibilities.

(1) The line of no contrast is always approximately perpendicular to $\boldsymbol{g}$, compare figure 4.7(a)–(c).

(2) Large black/white lobes are observed when the defects are near the surface, figure 4.7(a) at A, but there are no black/white depth oscillations.

(3) The size of most of the images is $\sim \xi_g$.

The above behaviour is typical of Ashby–Brown contrast from precipitates with an approximately

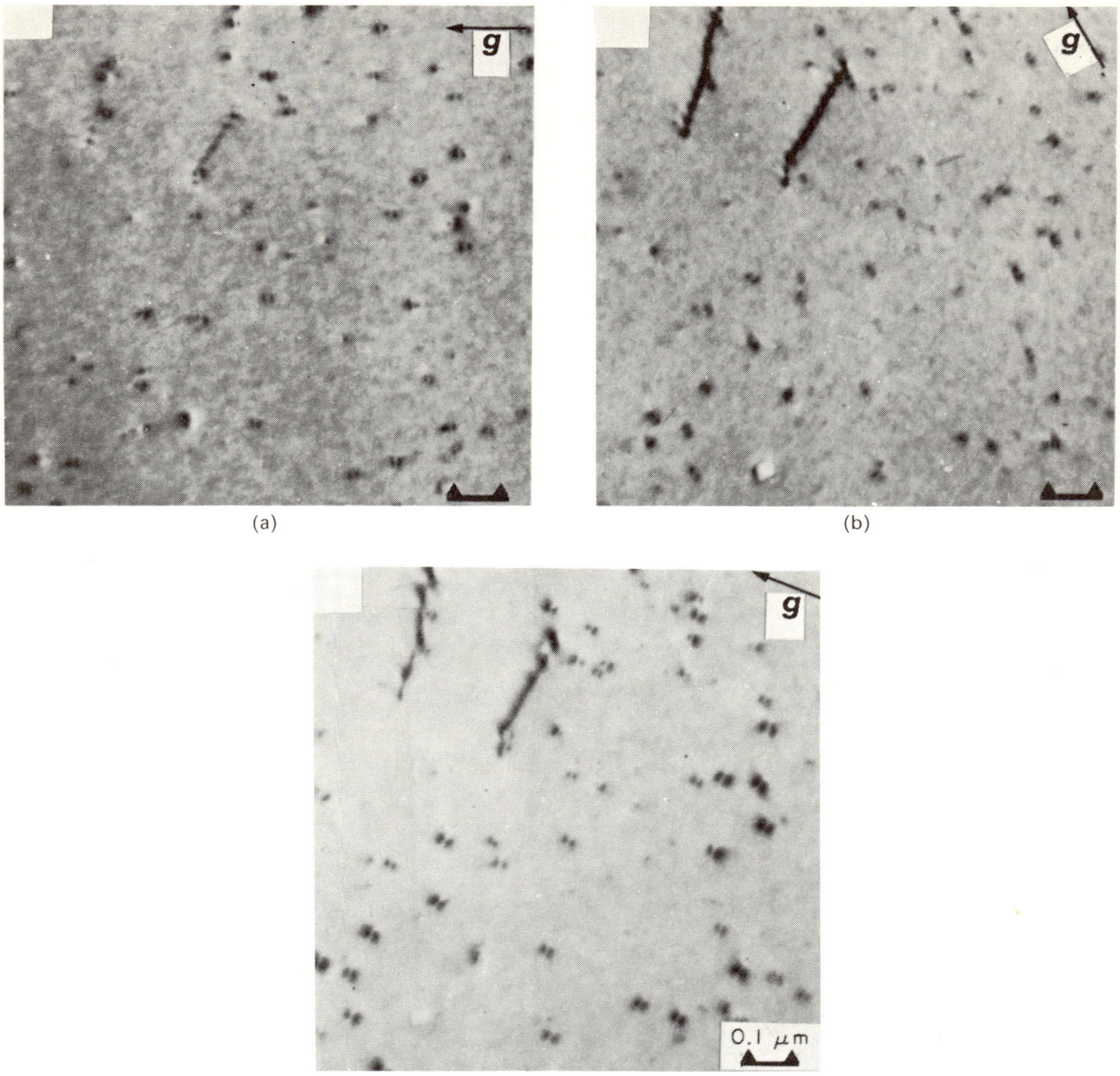

(a) (b) (c)

Figure 4.7 BF images illustrating the behaviour of defects 1 in quenched and aged stainless steel when imaged under different reflections: (a) $\boldsymbol{B} \sim [0\bar{1}1]$, $\boldsymbol{g}[1\bar{1}\bar{1}]$; (b) $\boldsymbol{B} \sim [0\bar{1}1]$, $\boldsymbol{g}[111]$; (c) $\boldsymbol{B} \sim [0\bar{1}1]$, $\boldsymbol{g}[0\bar{2}\bar{2}]$ [From Rowcliffe and Eyre (1971), courtesy of *Journal of Physics, F*]

spherical strain field with $P_s \gtrsim 0.2$ (section 3.11.1.2, table 4.5). They are not dislocation loops because the line of no contrast is always approximately perpendicular to $\boldsymbol{g}$ and large black/white lobes are obtained for defects near the surface. In fact, the defects were shown to be precipitates because they developed into small angular particles on further ageing.

Defects 2. The contrast behaviour of these defects under various strong two-beam diffraction conditions is shown in the BF images in figure 4.8(a)–(g), for which the operative reflections have been uniquely indexed by tilting within a given unit triangle, section 2.10.1.3. Three families of defect are present and examples are labelled A, B and C on the micrographs. The important aspects of the image behaviour are as follows.

(1) Defects A exhibit strong residual images with their line of no contrast always perpendicular to $\boldsymbol{g}$ for $\boldsymbol{g} \pm [200]$, $\boldsymbol{g} = \pm[020]$ and $\boldsymbol{g} = \pm[220]$, at orientations close to $\boldsymbol{B} \sim [001]$ (see figure 4.8(a)–(c)).

(2) Defects B and C are almost edge on when $\boldsymbol{B}[001]$ (see figure 4.8(a)–(c)).

(3) The residual images of defects B and C for $\boldsymbol{B} \sim [001]$ and $\boldsymbol{g} = \pm[200]$ or $\boldsymbol{g} = \pm[020]$ have a dot at each end. This is similar to residual images observed for edge-on prismatic dislocation loops (Maher and Eyre, 1971).

(4) The inside/outside image shifts obtained when $\pm\boldsymbol{g}$ experiments are carried out with $\pm[0\bar{2}2]$ and $\pm[\bar{3}11]$ are large and easily detectable, see figure 4.8(d)–(g). Again this is in agreement with results obtained from prismatic dislocation loops

for high-order reflections particularly when $\pm \boldsymbol{g} \cdot \boldsymbol{b} > 1$. This inside/outside contrast is summarised in table 4.7.

The above points (3) and (4) clearly show that the defects are dislocation loops and the visibility/invisibility behaviour, table 4.7, implies $\langle 100 \rangle$ Burgers' vectors. In the absence of detectable stacking fault contrast, it is concluded that $\boldsymbol{b}$ is $n\langle 100 \rangle$, where n is an integer. On the basis of $\pm \boldsymbol{g}$ experiments and the rules outlined in section 3.5.3 it is deduced that the loops are vacancy in sense.

However, one may obtain further information which indicates that the defects are more complicated than this analysis indicates. On further ageing at 700 °C these defects are replaced by CrP_3 platelets on $\{100\}$ planes with approximately the same density and distribution as the loops. Furthermore, $\{100\}$ habit planes for quenched-in loops in f.c.c. metals and alloys is most unusual, $\{111\}$ being almost always observed (Smallman,

Table 4.7 Summary of contrast behaviour of defects 2, figure 4.8

$\boldsymbol{B}$	001		$\bar{1}16$	133		112			
Defects	$\boldsymbol{g}$ = 020	$\bar{2}00$	220	$0\bar{2}\bar{2}$	$02\bar{2}$	$\bar{3}11$	$3\bar{1}\bar{1}$	$\boldsymbol{b}$	nature
A	out	out	out	inside	outside	inside	outside	$00\bar{1}$	vacancy
B	in	out	in	outside	inside	inside	outside	$0\bar{1}0$	vacancy
C	out	in	in	outside	outside	outside	inside	$\bar{1}00$	vacancy

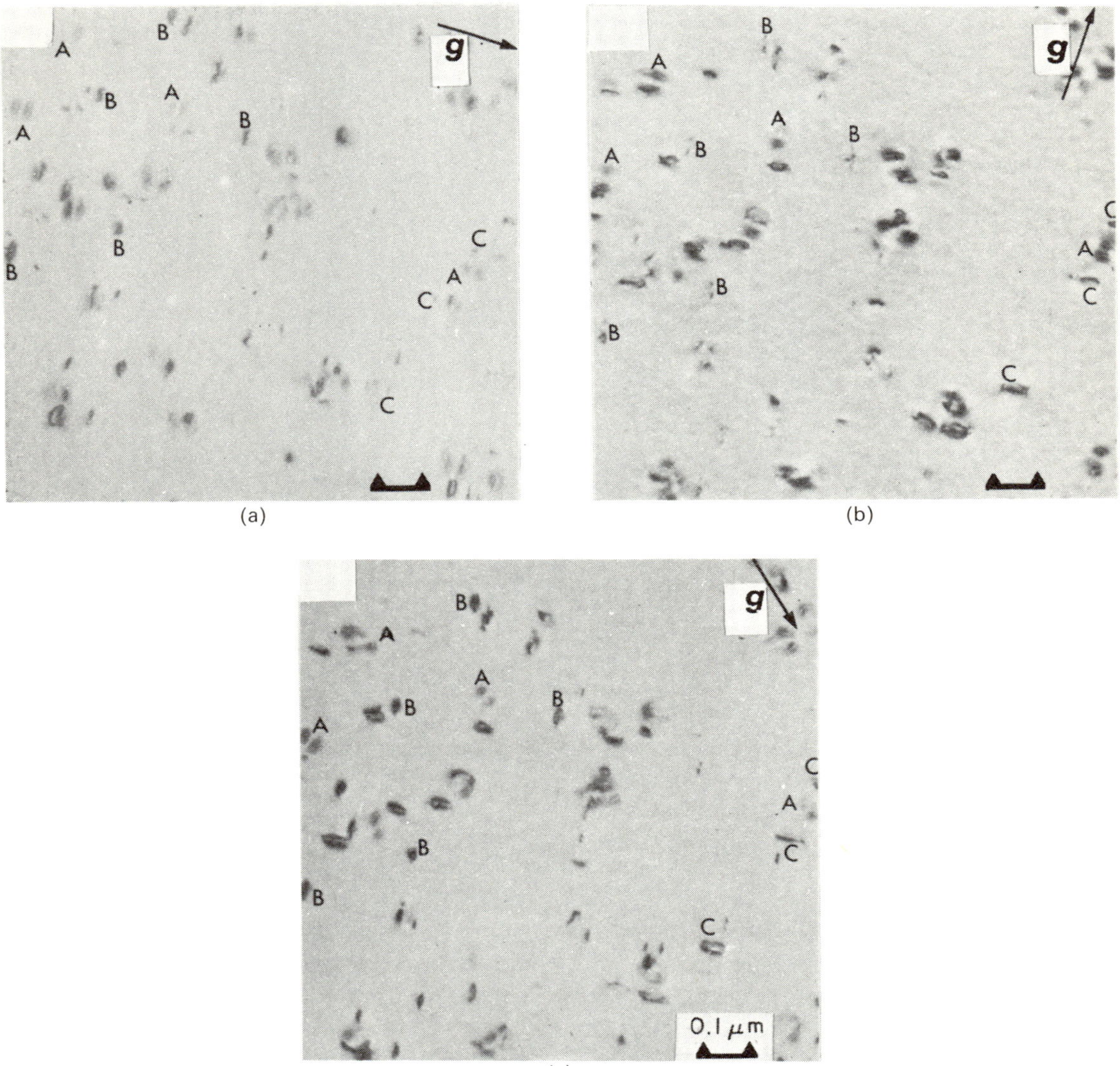

Figure 4.8 BF images of defects 2 in quenched and aged stainless steel for different reflections: (a) $\boldsymbol{B} \sim [001]$, $\boldsymbol{g}\,[020]$; (b) $\boldsymbol{B} \sim [001]$, $\boldsymbol{g}\,[\bar{2}00]$; (c) $\boldsymbol{B}\,[\bar{1}16]$, $\boldsymbol{g}\,[220]$; (d) $\boldsymbol{B}\,[133]$, $\boldsymbol{g}\,[0\bar{2}\bar{2}]$; (e) $\boldsymbol{B}\,[133]$, $\boldsymbol{g}\,[02\bar{2}]$; (f) $\boldsymbol{B}\,[112]$, $\boldsymbol{g}\,[3\bar{1}\bar{1}]$; (g) $\boldsymbol{B}\,[112]$, $\boldsymbol{g}\,[\bar{3}11]$ [From Rowcliffe and Eyre (1971), courtesy of *Journal of Physics, F*]

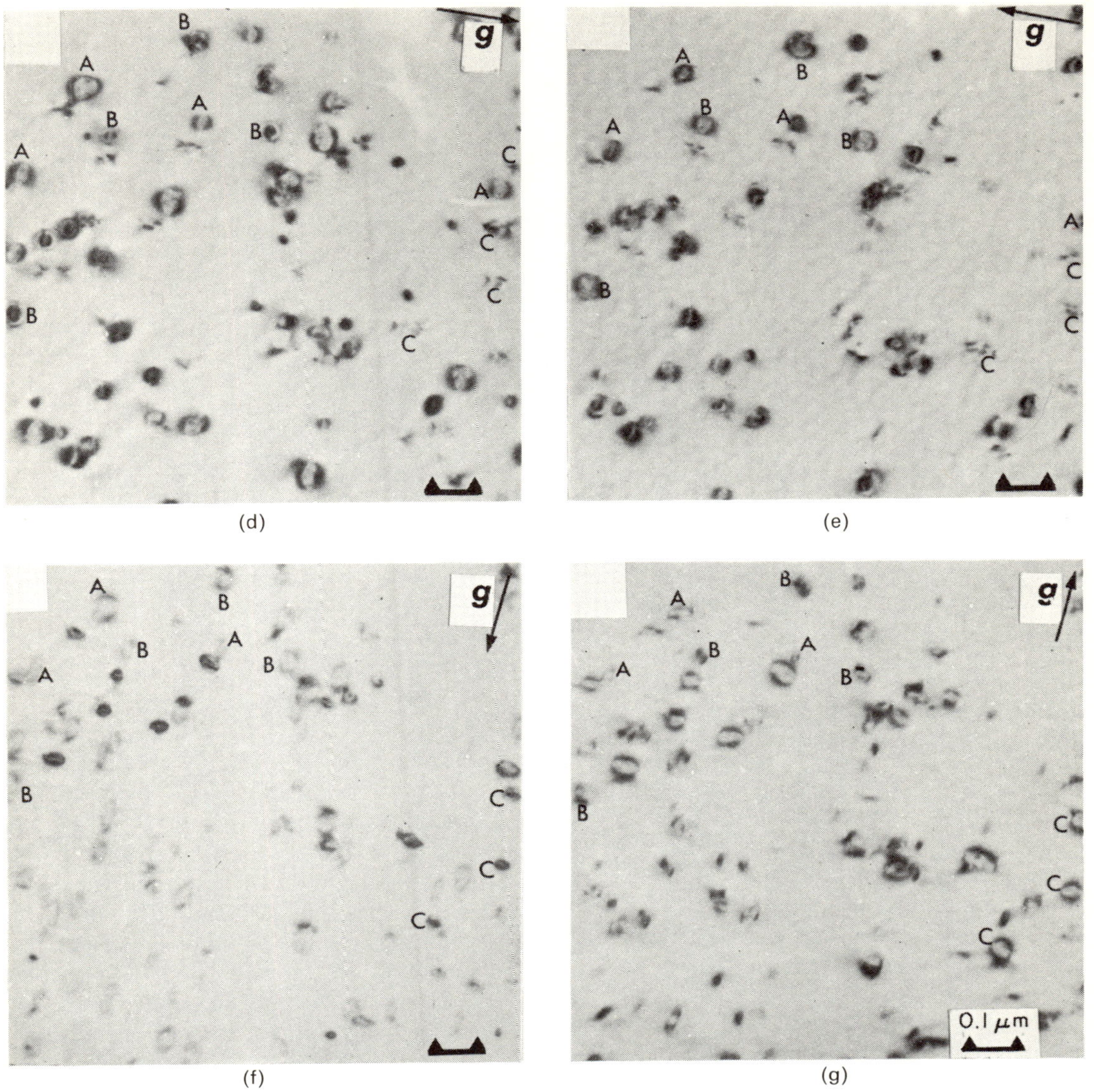

Figure 4.8 (*continued*)

1970). Therefore, it is in fact most likely that these defects consist of vacancy/solute complexes and are not simply quenched-in vacancy loops.

4.4 Characterisation of Voids, Strain-free Precipitates and Etch Pits

Characterisation of these defects, purely on the basis of contrast behaviour, is not yet possible. For the particular case of voids versus precipitates, additional metallurgical information is always required and the use of CDF using precipitate reflections is essential if they are visible in SADP. The image characteristics of the above three defects are summarised in table 4.8.

These defects may be readily distinguished from those described in the previous section because of the absence of any strain field image. Furthermore, etch pits, which are confined to the surface of the specimen, should be readily detectable using stereo microscopy.

4.4.1 *Identification of Voids in Stainless Steel* (*after B. L. Eyre*)

Shown below is a typical example of the differentiation between voids and precipitates for the high-phosphorous steel described in section 4.3.2.1. In this case the material had been solution treated at 1300 °C for 15 min, aged at 500 °C for 50 hours and the defects shown in figure 4.9 were observed. The images exhibited bright contrast in dark thickness fringes and dark contrast in bright thickness fringes which is characteristic of either void contrast or strain-free precipitates with low structure factor contrast. Confirmation that they were not surface pits was obtained by stereo microscopy. These defects were stable on annealing, which is not

Table 4.8 Crystal defects giving thickness contrast, that is without significant strain field

Defect	Type of image	Visibility, BF	Characteristics, BF	Quantitative information	Reference
cavity, that is negligible strain field	black or white dot depending on thickness if s small. Black or white dot underfocus respectively if s large	s small—easily visible near thickness fringes; s large—generally visible	contrast arises essentially from the thickness difference (thickness contrast). Most important characteristic is change through focus from bright to dark using s large	size and shape of cavities $\gtrsim 100$ Å diameter may be determined. Density may be measured. However, very small cavities $\lesssim 50$ Å may only really be detected not measured. Minimum detectable size ~ 10 Å	section 3.12
precipitate	contrast very similar to that of a cavity would be expected if the precipitate does not diffract strongly. However, the through-focus behaviour has not been checked specifically. Note that, if the precipitate is crystalline, tilting should cause it to diffract and rapidly become dark and then bright for a very small angle of tilt. Furthermore, CDF or fine structure of SADP should indicate presence of precipitate				
etch pit	should exhibit the same thickness contrast as cavity but contrast has not been studied in detail. Should be most readily detected using stereo microscopy to demonstrate that it lies at the surface				

characteristic of precipitates which would be expected to coarsen but is observed for voids (Westmacott *et al.*, 1968).

4.5 Conditions for the Best Visibility and Quantitative Analysis of Different Types of Defect

The best conditions for visibility of the different types of defect in BF under two-beam conditions are listed in tables 4.1, 4.4, 4.5 and 4.8. They may be summarised as follows.

$s = 0$ ($w = 0$) gives best images of stacking faults, plate precipitates with displacement contrast, antiphase domain boundaries, while $s_1 = -s_2$ for δ boundaries give best images.

$s > 0$ that is positive ($w > 0$) but small gives best images for strain field contrast from dislocations,

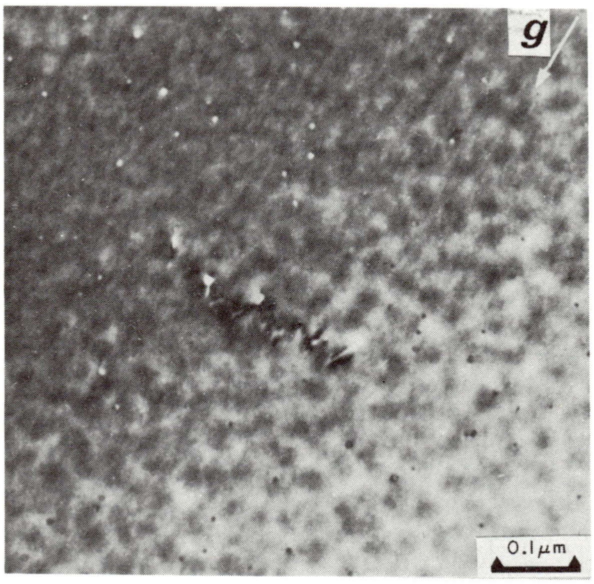

Figure 4.9 A CDF image of defects 3 in quenched and aged stainless steel: $\boldsymbol{B} \sim [1\bar{1}0]$, $\boldsymbol{g}\,[11\bar{1}]$

gas bubbles, dislocation loops, misfitting precipitates, etc.

$s = 0$ or $s > 0$ ($w = 0$ or $w > 0$) give good images for voids and small particles imaged as a result of structure factor contrast in or near thickness fringes.

Best penetration is usually obtained in BF with $w \approx 0.5$–1.0, see section 3.2.3.2 and figure 3.3(b). Consequently, in practice, most images are formed under these conditions because defects are best studied in the thicker regions of the foil where thickness fringes do not complicate the image and the defect structure is more typical of the bulk material. Note, however, that $s = 0$ conditions should be used when determining the sense of stacking faults or precipitate plate displacement so that the nature of the outer fringes may be readily determined. Note also the values of w should be selected in the SADP not using the image as a guide to best overall image quality because different defects give best contrast with marginally different values of w. Thus dislocations show different contrast for a given value of w depending on $\boldsymbol{g}\cdot\boldsymbol{b} = 1$ or 2, see section 3.4.3.1 (Loretto and France, 1969).

Two cases of special difficulty can be defined below.

4.5.1 *High Density of Defects of the Same Type*

If the defects are imaged by strain fields, then we have the following.

(1) The weak beam technique may be used to obtain a more detailed picture of, for example, dislocation tangles or networks, see figures 3.53(d),(e), 3.39(a),(b).

(2) Use CDF imaging with precipitate reflections where possible. This approach is particularly useful when a high density of precipitates of different

crystal structure are present and the strain field images are complex. An example of the use of this approach in an Al–Zn–Mg alloy is given in section 4.9.2.

(3) Quantitative information such as the dislocation Burgers vector, the sense and direction of strain field, etc., cannot be obtained from the strain field image where overlap of images occurs. Under these conditions regions of low defect density must be used for analysis.

If the defects are imaged by thickness effects (voids) or structure factor effects (precipitates), then thinner regions of the foil may be used to avoid image overlap.

4.5.2 *Defects of Different Types Present Simultaneously*

If the defects are imaged using strain fields, then the following occurs.

(1) Considerable difficulty will be experienced if they have the same type of image, for example misfitting precipitates and gas bubbles present. The most useful technique in such cases is CDF using precipitate reflections. Quantitative work still requires regions of low defect density.

(2) When the type of image is different, for example dislocations and misfitting precipitates, the weak beam technique will enable more detail to be seen, but quantitative work requires simple defect configurations in low defect density regions of the specimen.

(3) In special cases images may be separated if they exhibit best contrast under different diffracting conditions. The best example of this are voids or precipitates imaged by structure factor contrast in the presence of strain field images like dislocations. Under these conditions it is best to obtain BF images with two-beam conditions on a higher-order reflection with w large, for example $\{330\}$ in b.c.c. crystal structures. The dislocation images are faint and do not obscure the void contrast. The voids may then be identified from their through-focus properties.

III. DIFFERENTIATION BETWEEN SIMILAR EFFECTS IN DIFFRACTION PATTERNS

In a given diffraction pattern, additional fine structure may be present arising from a wide range of crystal defects, see sections 2.15–19.

Two types of effect occur, namely, (1) additional spots or satellites and (2) streaks crossing the diffraction pattern. In general it is not possible to identify crystal defects from the diffraction patterns alone because a wide range of crystal defects may be responsible for (1) or (2) above.

In order to interpret the fine structure in a diffraction pattern and obtain useful information the following type of data may be required.

(1) The influence of small amounts of specimen tilt ($\pm \sim 3°$) about several axes in the SADP, on the details of the pattern.

(2) The influence of large tilts (20–60°) to produce ***B*** parallel to several crystallographically simple zone axes, for example [100], [011], [111] in cubic crystal structures. This enables a map of reciprocal space to be constructed.

(3) A large number of CDF images obtained using the features of interest in the diffraction pattern to determine which crystal defects produce them.

In complex microstructures, such as those of many commercial materials, a number of different crystal defects may lie within the selected area aperture and thus contribute to the SADP. Consequently, very complex diffraction patterns may occur and (1)–(3) above must all be used to identify all features of the SADP.

4.6 Characterisation of Extra Spots and Satellites

Extra spots may arise in diffraction patterns for a number of reasons which are listed in table 4.9. It is not possible to formulate a specific series of rules to follow in interpreting the fine structure of diffraction patterns because of the diverse range of crystal defects that may be responsible. However, by combining the information obtained from points (1)–(3) above with the overall defect structure of the specimen obtained from the BF image, considerable quantitative information may be obtained.

Examples of diffraction pattern analysis associated with specific crystal defects are included in other sections. Thus twins are discussed in section 2.12; modulated structures in 2.17.2, 4.10 long-range order in sections 2.16.2, 4.14.1; precipitates in sections 2.11.1, 2.13, 4.9.1, 4.9.2, 4.11.1, etc. In addition, the importance of double diffraction is stressed in several examples (sections 2.16.1, 4.9.2, 4.13.1). In all cases identification of the origin of the additional spots is based on the features described in table 4.9.

4.7 Characterisation of Streaks

Streaks may arise in diffraction patterns for a number of reasons which are listed in table 4.10. In this case, it is not possible to list a specific series of rules to be followed in investigating such effects because a wide range of crystal defects could be

Table 4.9 Characteristics of extra spots in diffraction patterns arising from different crystal defects

Crystal defect	Characteristic distribution of extra spots	Experimental conditions under which the effect is visible	Method by which the defect producing the effect may be identified	Quantitative information obtainable	Reference
precipitate (coherent or semi-coherent)	a regular array characteristic of the crystal structure of the precipitate	most useful information obtained with ***B*** close to a series of simple, low-index matrix zone axes with known tilts between	use CDF with precipitate reflections to identify which variants are responsible for particular spots. Index spots for various ***B*** and check crystal structure with x-ray data files	identification of crystal structure, determination of precipitate particle size, density and shape	section 2.11.1
precipitate (incoherent)	apparently random spots in addition to matrix pattern	any ***B*** in matrix	for large precipitates obtain a series of SADP from individual particles with simple low-index ***B***	identification of crystal structure	section 2.13
twins, one large or several with same mode	a regular array characteristic of twin mode	as first entry in this column above including twin and matrix in selected area aperture	CDF may be used to identify twin spots which may then be indexed	twin mode	section 2.12
twins, small different systems	a complex series of regular arrays of spots	as first entry in this column above including twin and matrix in selected area aperture	as third entry in this column, but noting which twins are responsible for particular arrays	twin modes but may be difficult	section 2.12
modulated structures (a) spinodal	pairs of satellites equally spaced about diffracted spots. Line joining satellites parallel to direction of modulation, usually ⟨100⟩ f.c.c.	***B*** exactly parallel to zone axes and perpendicular to direction of modulation	CDF using satellites only and influence of ageing on satellite spacing	wavelength and direction of modulation	section 2.17.2.1
(b) regular arrays of APB	arrays of satellites around diffracted beams. Details depend upon particular geometry of crystal structure and APB	***B*** exactly parallel to low-index zone axes. Double diffraction can produce extra spots	BF or CDF using strong superlattice reflections make APB visible	spacing of APB	section 2.17.2.2
(c) regular arrays of dislocations	as for (b) above, satellite row parallel to dislocation row	***B*** exactly parallel to low-index zone axes. Double diffraction can produce extra spots	conventional BF or WB images for high dislocation densities	spacing of dislocations from image	section 2.17.2.3
long-range order	regular arrays of extra spots, spacing simple fraction of matrix spots	as for first entry in this column	indexing of SADP, CDF using superlattice reflections to show APB, etc., or careful imaging to show δ boundaries	identification of ordered crystal structure	section 2.16.2
double diffraction	either clusters of satellites around transmitted and diffracted spots, reflecting symmetry of spot pattern, or additional spots at simple fractional distances from main spots and not accounted for by structure factor	***B*** very close to zone axis such that the full array of spots is visible	can be identified (a) by tilting slowly away from exact ***B*** when spots should disappear rapidly, (b) by checking in other zone axes or (c) by CDF images which may also be useful because they outline interface at which double diffraction occurs	aids in solution of SADP when identified	section 2.16.1

Table 4.9 (*continued*)

Crystal defect	Characteristic distribution of extra spots	Experimental conditions under which the effect is visible	Method by which the defect producing the effect may be identified	Quantitative information obtainable	Reference
Ewald sphere intersects streaks $s \neq 0$	pairs or clusters of satellites around diffracted spots. Only present if $s \neq 0$ and spacing increases as s increases	any conditions where a specific reflection is excited, except $s = 0$, provided defects are present that produce streaks in reciprocal space	can be used to identify defect responsible, and behaviour of spots on changing s, that is reduction in spacing	spacing of satellites can be used to determine plane of defect	sections 2.18.1, 2.17

responsible. However, combination of the information contained in points (1)–(3) above with the defect morphology obtained from the BF image should provide enough information for the effects to be explained.

Examples of analysis of streak effects are discussed in other sections. Thus planar precipitates are discussed in sections 2.18.1, 4.9.2; planar defects such as stacking faults in sections 2.17.1, 4.2.1.1; APB in section 2.17.1; δ boundaries in sections 2.17.1, 4.13.3; grain boundaries in section 2.17.1; rod-shaped precipitates in section 2.18.1; short-range order in section 2.19.2; elastic strains near precipitates in section 2.18.2; sheets and discs of intensity in sections 2.18.1, 2.19. In all cases identification of the defects responsible is based on the characteristics summarised in table 4.10.

Table 4.10 Characteristics of streaks in diffraction patterns arising from different crystal defects

Crystal defect	Characteristic direction of streaks	Experimental conditions under which the effect is visible	Method by which the defect producing the streak may be identified	Quantitative information available	Reference
shape effects (a) planar defects such as precipitates, twins	streaks run perpendicular to habit plane	$s = 0$ for reflections with the defect plane normal exactly in the plane of the diffraction pattern. Usually simple low-index $\boldsymbol{B}$ (applies to (a) and (b))	use CDF on streak and satellites arising from cutting streaks $s \neq 0$, see table 4.9 (applies to (a) and (b))	shape and habit plane of defect (applies to (a), (b) and (c))	section 2.18.1
(b) planar defects such as stacking faults APB, δ boundaries, grain boundaries	streaks run perpendicular to habit plane but the number of streaks depends upon the defect				section 2.17.1
(c) rod-shaped precipitates	streaks run perpendicular to long axis of rod. Streaks may be curved for non-rational $\boldsymbol{B}$	any value of s and $\boldsymbol{B}$ but most easily analysed at $s = 0$ with simple low-index $\boldsymbol{B}$	use CDF on streak		section 2.18.1
(d) short-range order SRO	diffuse streaks occur at positions that do not coincide with long-range order superlattice spots	exact $\boldsymbol{B}$ at simple zone axes of matrix	direct comparison with theory	there is still some discussion of exact interpretation	section 2.19.2
surface films	depends on particular case	any $\boldsymbol{B}$, $s = 0$	careful control of polishing conditions		section 2.19.1
elastic strain, for example around precipitate particles	streaks run in direction of maximum strain	$s = 0$ for direction of max. strain perpendicular to $\boldsymbol{B}$. Usually simple, low-index $\boldsymbol{B}$	as first entry in this column	direction of elastic strains if they can be separated from shape effects	section 2.18.2
intersecting of discs of intensity in reciprocal space	streaks present over a wide range of tilt	streaks occur for any rational or irrational $\boldsymbol{B}$	superimpose SADP for different $\boldsymbol{B}$ to map reciprocal space and use CDF on streaks to determine defects responsible	detection of early stages of ordering particularly some cases of short-range ordering in compounds	section 2.19

4.8 Conditions for Best Visibility and Analysis of Fine Structure in Diffraction Patterns

The best conditions for visibility of particular extra spots or streaks are included in tables 4.9, 4.10. They may be summarised as follows.

(1) Exact ***B*** parallel to low-index zone axes give most information for the identification of precipitates and the determination of their orientation relationships to the matrix; the identification of twin modes; studies of modulated structures; studies of streaks from precipitate, shape and elastic strain.

(2) Tilting so that ***B*** is close to exact zone axes is useful in studies of double diffraction effects; satellites arising from streaks produced by precipitates, stacking faults, etc.; maximisation of precipitate reflection intensities when the orientation relationship is not exact, or when several types of precipitate are present.

(3) Tilting through large angles to map reciprocal space and obtain a number of low-index exact ***B*** is useful for identification of precipitates; determination of orientation relationships; determination of the form of walls of diffuse intensity in the reciprocal lattice, determination of the distribution of shape streaks, etc., in the reciprocal lattice; determination of crystal structure for long-range order.

It is important to realise that, in order to record the information present in diffraction patterns, carefully focussed SADP must be used with the condenser 2 well overfocussed. A series of exposure times of up to two minutes may be necessary to record all the information in the diffraction pattern, particularly if several effects of different intensity such as faint streaks and strong satellites are present simultaneously because each may require a different exposure.

Difficulties will be encountered if several types of crystal defects are present simultaneously, as is frequently the case in commercial materials. Under such circumstances it is advantageous to try to simplify the microstructure if possible by changing the specimen treatment and then to use the information from the simple case as a basis for solving the more complex situation. In addition, extensive use of CDF images using all extra reflections from a given area of the specimen for a range of specimen tilts will be found to be an invaluable aid in solving the structure. In this context, a particularly useful example is shown in section 4.9.2. Here extensive use is also made of diffraction patterns calculated previously on the basis of likely precipitate structures and orientation relationships.

IV. EXAMPLES OF ANALYSES FROM METALS AND ALLOYS

4.9 Aluminium Alloys

A number of examples of different studies of precipitation hardening alloy systems are described in this section.

4.9.1 *Determination of the Orientation Relationship Between η Precipitate and Matrix in an Al–6.35 wt% Zn–3.15 wt% Mg Alloy (after K. V. Subbaraô)*

There have been twelve different precipitates detected in this system within the narrow range of alloys 5–7 wt% Zn and 0.5–3 wt% Mg. These are as follows: the 'η′' phase (one type); the 'T' phase (one type); the 'X' phase (one type); the η phase (nine types).

The object here is to show the orientation relationship between large η particles and matrix using spot patterns obtained from single particles. The approach used was the following.

(1) To over-age to produce coarse precipitates so that individual rather than groups of particles could be studied. This is particularly important in this alloy because a number of different precipitates may be present simultaneously.

(2) To obtain several different SADP that include both precipitate and matrix spots for different simple ***B*** in the matrix. Limited specimen tilt was available; consequently different matrix grains were used to obtain these different ***B***.

Figure 4.10(a) shows a typical coarse lath array produced by ageing at 180 °C for 65 hours after direct quenching from the solution treatment temperature of 465 °C. Figure 4.10(b) shows the SADP from figure 4.10(a).

Additional examples of SADP with different matrix ***B*** and containing both precipitate and matrix reflections are shown in figure 4.11(a),(c).

Table 4.11 Interplanar spacings for η with $a = 5.21$ Å and $c = 8.60$ Å

hkl	*d* (Å)	*hkl*	*d* (Å)
100	4.518	002	4.303
110	2.601	102	3.116
200	2.259	112	2.230
210	1.707	202	2.000
220	1.304	004	2.152
300	1.506	104	1.943
310	1.253	114	1.660
101	4.000	204	1.558
111	2.496	006	1.435
201	2.185	008	1.076

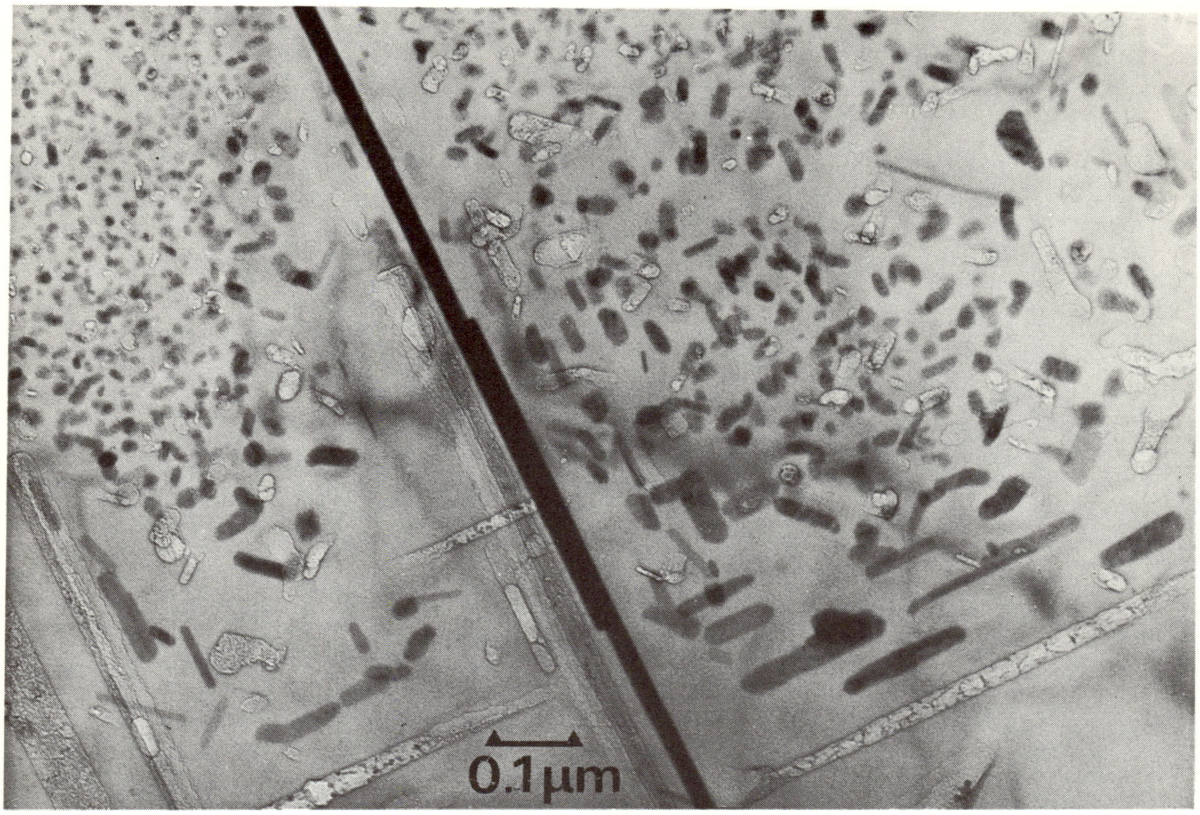

(a)

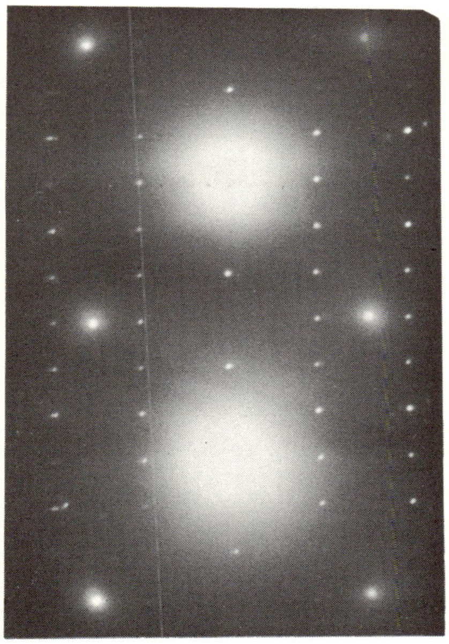

(b)

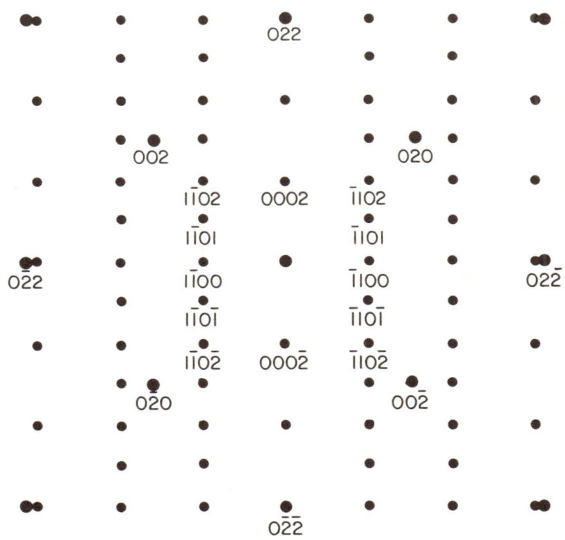

(c)

Figure 4.10 (a) A long η lath in quenched and aged Al–6.35 wt % Zn–3.15 wt % Mg alloy; (b) the SADP from (a); (c) the indexed pattern [Courtesy of K. V. Subbaraô]

Table 4.11 contains the interplanar spacings for the η phase for $a = 5.21$ Å and $c = 8.60$ Å (Gjønnes and Simensen, 1970). Using these values and the camera constant determined from the matrix reflections, the SADP can be indexed as shown in figures 4.10(c), 4.11(b),(d). The orientation relationships may be obtained by assuming that the zone axes are parallel in figures 4.10, 4.11 and that rows of parallel spots in the two phases correspond to parallel planes. Thus superimposed matrix and η stereograms describing the indexed diffraction patterns in figures 4.10, 4.11 may be constructed as outlined in section 2.11.1, see figure 4.12. This is a particularly simple case because all the planes

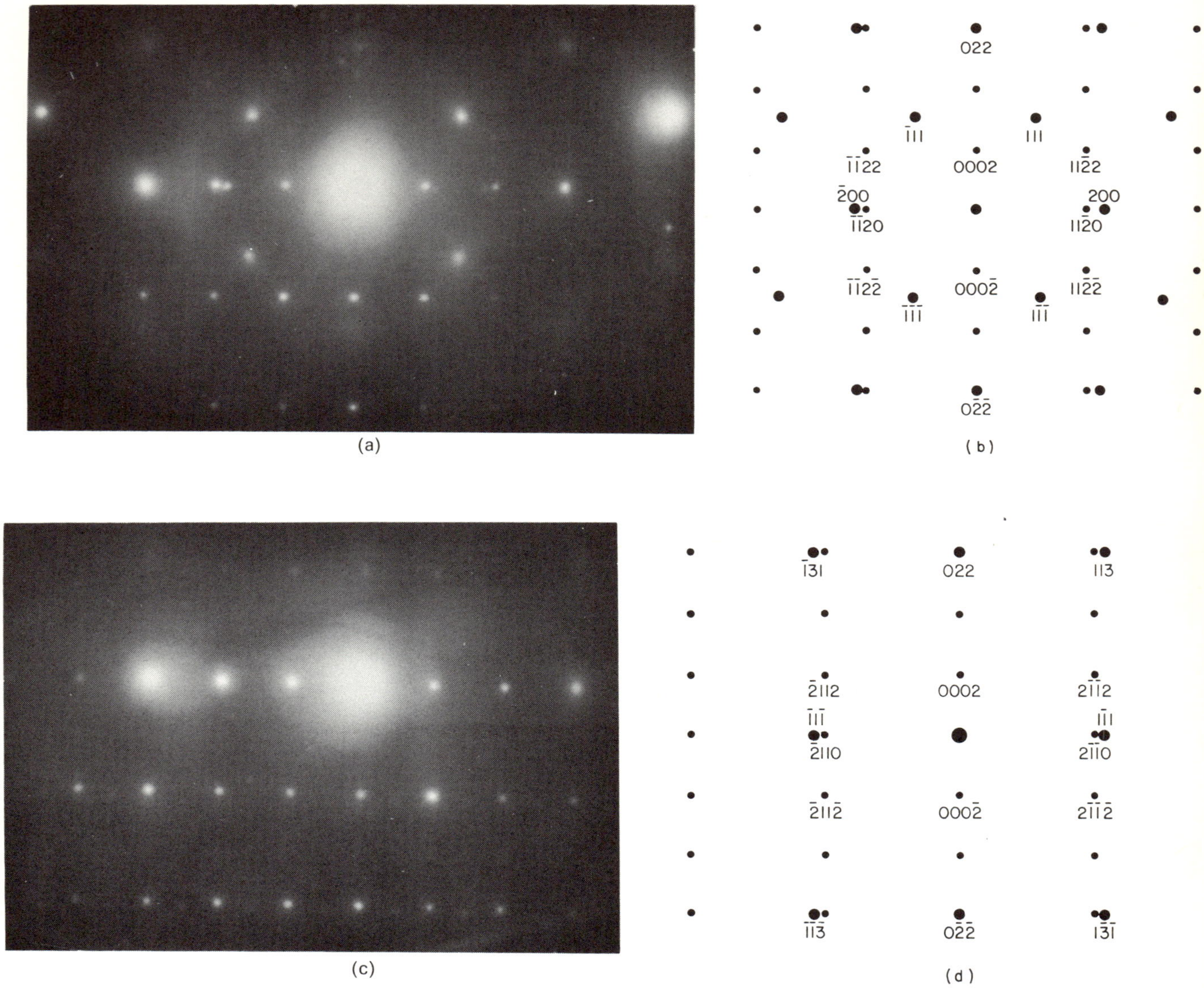

Figure 4.11 (a), (c) SADP from a large η lath in Al–6.35 wt% Zn–3.15 wt% Mg and the indexed patterns: (b) $\boldsymbol{B} \sim [01\bar{1}]_{matrix}//[\bar{1}100]_\eta$ and $(0001)_\eta//(011)_{Al}$, $(11\bar{2}0)_\eta//(100)_{Al}$; (d) $\boldsymbol{B} \sim [21\bar{1}]_{matrix}//[01\bar{1}0]_\eta$ and $(0001)_\eta//(011)_{Al}$, $(2\bar{1}\bar{1}0)_\eta//(1\bar{1}1)_{Al}$ [Courtesy of K. V. Subbaraô]

in the hexagonal phase of interest in this analysis have the same indices as their normals, that is $l = 0$, see appendix 2. We therefore use standard (0001) hexagonal and (011) cubic stereograms to represent the orientation relationships, as shown in figure 4.12. The orientation relationship obtained from the analysis is

$$(11\bar{2}0)_\eta//(100)_{Al} : [0001]_\eta//[011]_{Al}$$

or

$$(\bar{1}100)_\eta//(01\bar{1})_{Al} : [0001]_\eta//[011]_{Al}$$

which corresponds to that proposed by Embury and Nicholson (1965) and is η_9 on the classification scheme of Gjønnes and Simensen (1970).

4.9.2 *Confirmation of the Presence of Two Precipitates η' and η_1 in an Al–5.77 wt% Zn–1.08 wt% Mg Alloy (after C. J. Simensen and A. J. de Ardo, Jr.)*

Although a large number of different precipitates can form in aged Al–Zn–Mg alloys, see section 4.9.1, for those ageing treatments used commercially the most common precipitates are η', η_1, η_2 and η_4. In this section we shall show how the η' and the η_1 phases may be detected when they occur simultaneously together with η_2. The crystal structure, morphologies, lattice parameter and orientation relationships of these phases are summarised in table 4.12.

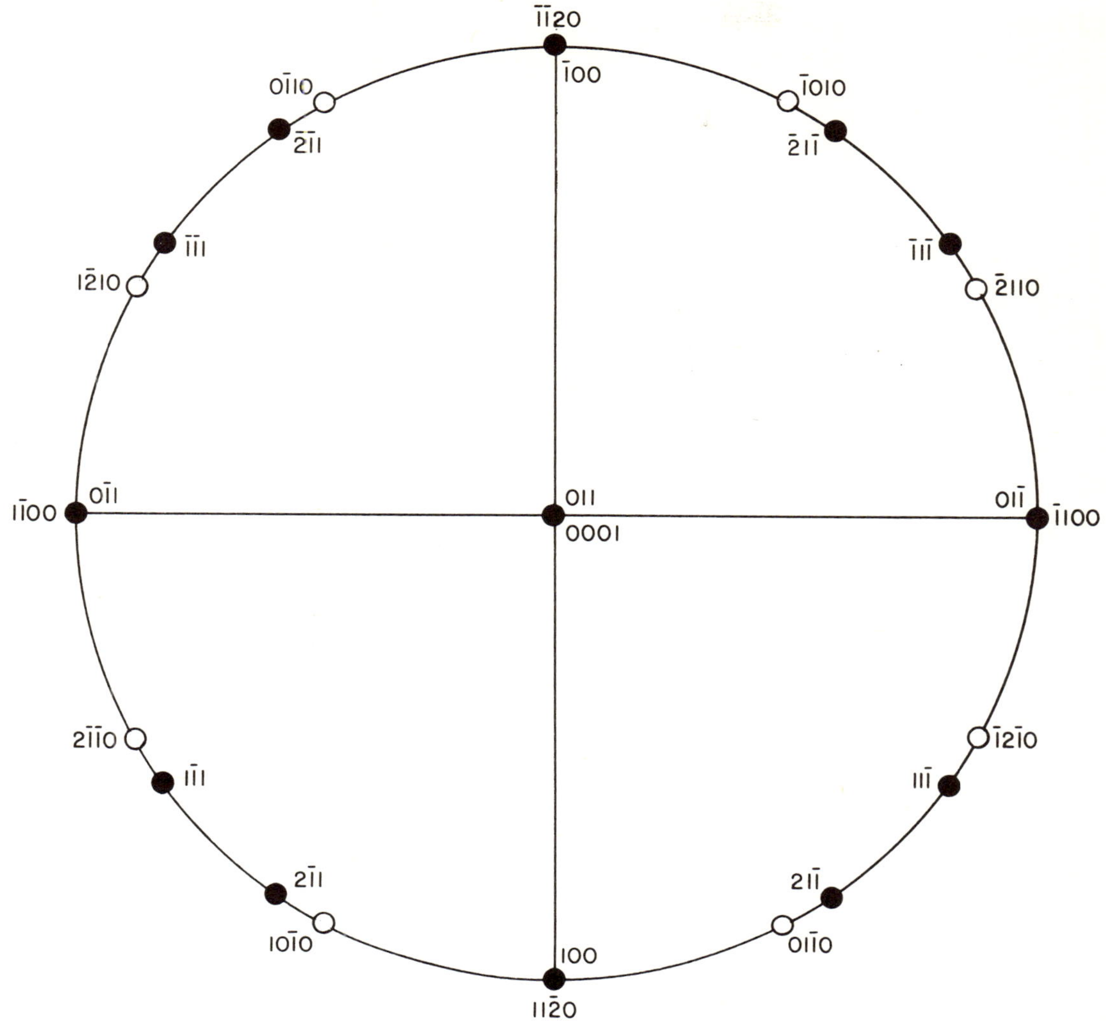

Figure 4.12 A stereographic representation of the orientation relationship between the η precipitate and aluminium matrix $(11\bar{2}0)_\eta//(100)_{Al}$; $[0001]_\eta//[011]_{Al}$ [Courtesy of K. V. Subbaraô]

Because of the large number of different variants of the orientation relationships, experimental SADP are complex. In such situations the best approach (section 2.11.1) is to compute the form of the diffraction pattern to be expected for the known (or likely) orientation relationships for the exact ***B*** in the matrix obtained in a series of experimental patterns. ***B*** may be determined from the positions of Kikuchi lines or considerations of spot brightness, sections 2.7.3, 2.7.2.1. All variants of the orientation relationship should be included in the computation and the computed pattern may then be compared in detail with the experimental one. This approach forms the basis of the example given here.

The specimen is an Al–5.77 wt% Zn–1.08 wt%

Table 4.12 Crystallographic data for the η' and η_1 phases in Al–Zn–Mg alloys

Precipitate type	Precipitate shape	Crystal structure	Lattice parameter (Å)[a] a	Lattice parameter (Å)[a] c[b]	Orientation relationships data	References
η'	plates on $\{111\}_{Al}$	hex.	4.97	5.54	$(10\bar{1}0)//(110)_{Al}$; $(0001)//(1\bar{1}\bar{1})_{Al}$	Gjønnes and Simensen (1970)
η_1	plates on $\{100\}_{Al}$	hex.	5.21	8.60	$(10\bar{1}0)//(001)_{Al}$; $(0001)//(110)_{Al}$	Schmalzreid and Gerold (1958)
η_2	plates on $\{111\}_{Al}$	hex.	5.21	8.60	$(10\bar{1}0)//(110)_{Al}$; $(0001)//(1\bar{1}\bar{1})_{Al}$	Gjønnes and Simensen (1970)

[a] Gjønnes and Simensen (1970).
[b] The value of c has been recently reported as 13.74 Å (Auld and Cousland, 1971). If this is correct, the conclusions of this section will still be valid but the l index of $(hkil)$ planes must be doubled.

Mg alloy annealed for 1 hour at 485 °C, water quenched to 20 °C for 5 s, then aged at 150 °C for 28 hours followed by water quenching. Figure 4.13(a) is a typical BF image and figures 4.13(b),(c) show SADPs obtained from this area with ***B*** [1$\bar{1}\bar{1}$] and [1$\bar{1}\bar{2}$] in the matrix, that is the specimen was tilted about a [220] axis (common reflection). Figure 4.14 shows the computed diffraction patterns for the lattice parameters and orientation relationships in table 4.12, η' and η_1 spots being denoted by × and □ respectively. Comparison of the patterns shows good agreement, although

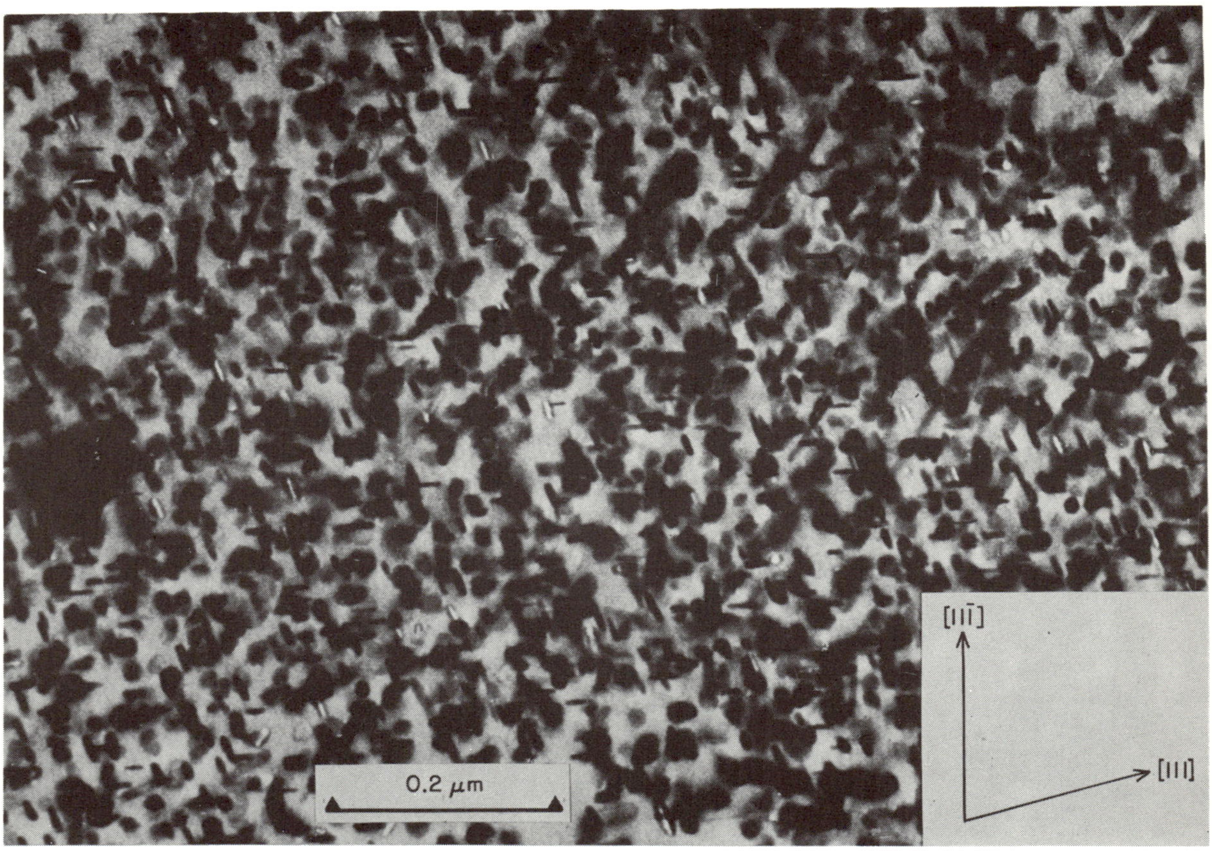

(a)

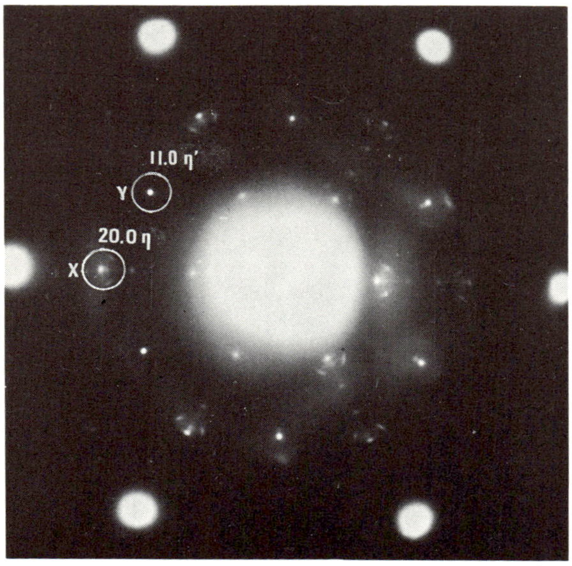

(b)

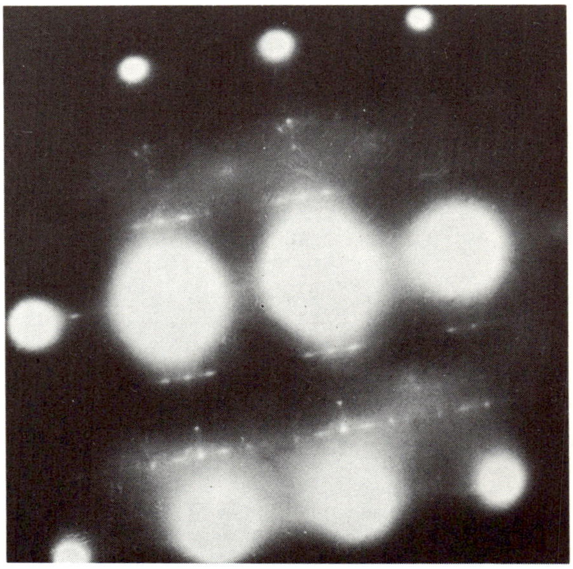

(c)

Figure 4.13 (a) A BF image showing strain field contrast in Al–5.77 wt % Zn–1.08 wt % Mg aged at 130 °C for 28 hours. [Courtesy of C. J. Simensen] (b) SADP which shows ***B***[1$\bar{1}\bar{1}$]$_{Al}$. (c) SADP which shows ***B***[1$\bar{1}\bar{2}$] [From Gjønnes and Simensen (1970), courtesy of *Acta Metallurgica*]

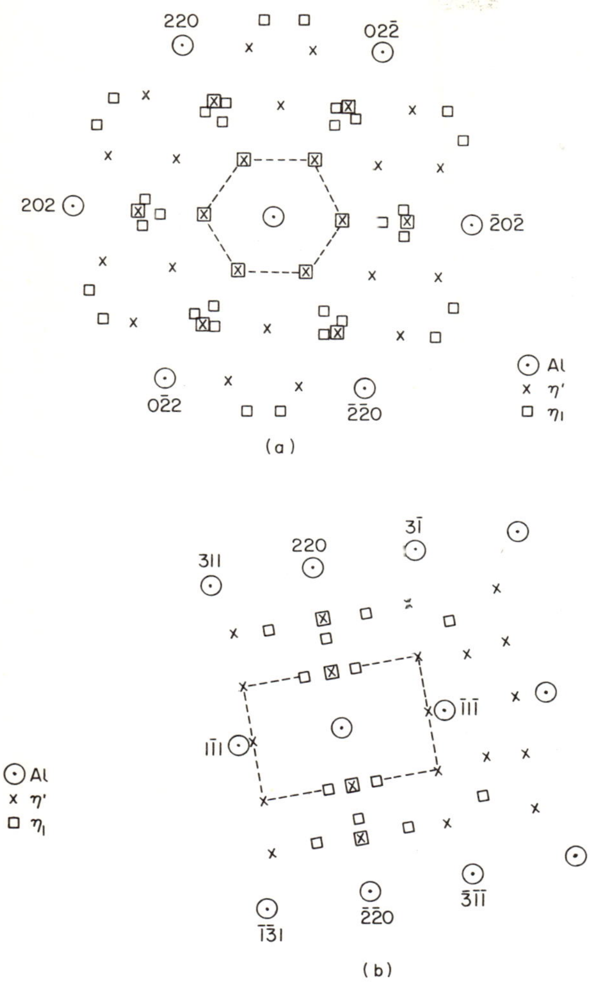

Figure 4.14 Computer-generated SADP for the η' and η_1 orientation relationships in table 4.12 [Courtesy of C. J. Simensen]

there are additional spots present in each pattern that are not predicted. These arise for three reasons.

(1) Double diffraction (section 2.16.1) which accounts for several features. One of the clearest examples is the repeat in figure 4.13(b) of the precipitate reflections predicted between the centre spot and the (220) reflection between the ($\bar{1}1\bar{1}$) and (13$\bar{1}$) reflections. A second example is the repeat in figure 4.14(a) of the predicted four-spot cluster between the centre spot and the (02$\bar{2}$) reflection at the inner η' reflection along the same direction.

(2) There is η_2 phase present, see table 4.12 for crystallographic data.

(3) The spots are streaked in ⟨111⟩ and projected ⟨111⟩ because η' particles produce strong streaks in these directions.

CDF images are particularly useful to test the validity of the above interpretation and to measure the particle sizes, shapes and densities of the individual phases. Figure 4.15(a) shows CDF on the η' reflections circled in figure 4.13(b) and delineates one variant of plates on {111}. However, reference to table 4.12 shows that η' and η_2 precipitates are *both* plates on $\{111\}_{Al}$ and computation shows that these also produce spots within the region of the diffraction pattern selected to produce figure 4.15(a). Consequently, figure 4.15(a) could contain both η' and η_2 particles. However, the η_2 precipitates may be distinguished from η' because they give characteristic Moiré patterns for CDF in position Y, figure 4.13(b). Alternatively, the η' precipitates may also be distinguished by slowly tilting the specimen away from the $[1\bar{1}\bar{1}]_{Al}$ zone when the η' CDF contrast persists compared with η_2 because of the extensive streaking of η' reflections in [0001] (Gjønnes and Simensen, 1970). Using the former approach the precipitates near A, figure 4.15(b), may be identified as η_2 precipitates. Clearly, on the basis of the above discussion η', η_1 and η_2 precipitates are present in figure 4.13(a).

The Moiré patterns in figure 4.15(b),(c) may be interpreted in terms of interference between $\{11\bar{2}0\}$ η_2 reflections with five double diffracted spots nearby (Gjønnes and Simensen, 1970). Shifts in the Moiré patterns, see arrows in figure 4.15(c), may be interpreted in terms of stacking faults in the η_2 precipitate (Gjønnes and Simensen, 1970).

The advantages of using computed diffraction patterns are as follows.

(1) The exact ***B*** may be computed, even if it is not exactly parallel to a simple zone axis.

(2) The exact positions of the spots are obtained.

The alternative method is to construct the diffraction patterns directly from the stereograms. The details of such an approach for the same problem are given below for ***B*** $\langle 112\rangle_{Al}$. The necessary crystallographic manipulations are described in appendix 1.

Consider first the case of the η' precipitate. Figures 4.16(a),(b) show the stereographic projections which if superimposed would describe the orientation relationship in table 4.12. Only the projections of planes on the edge of the stereograms, that is those of interest in terms of the SADP, are included. The ***B*** of interest is $[1\bar{1}2]$ which is marked in the ($1\bar{1}\bar{1}$) stereogram in figure 4.16(b). The great circle corresponding to this pole contains all the precipitate reflections for ***B*** $[1\bar{1}\bar{2}]$ for the $(0001)_{\eta'}//(1\bar{1}\bar{1})_{Al}$ variant. In fact, this great circle only includes the (10$\bar{1}$0) planes which are parallel to $(110)_{Al}$ and are drawn in figure 4.16(c).

4.9.2.1 *The* ($1\bar{1}1$) *variant*

If the $[1\bar{1}1]$ pole is moved to the centre of the stereogram by a rotation of 70° 36′ about [110],

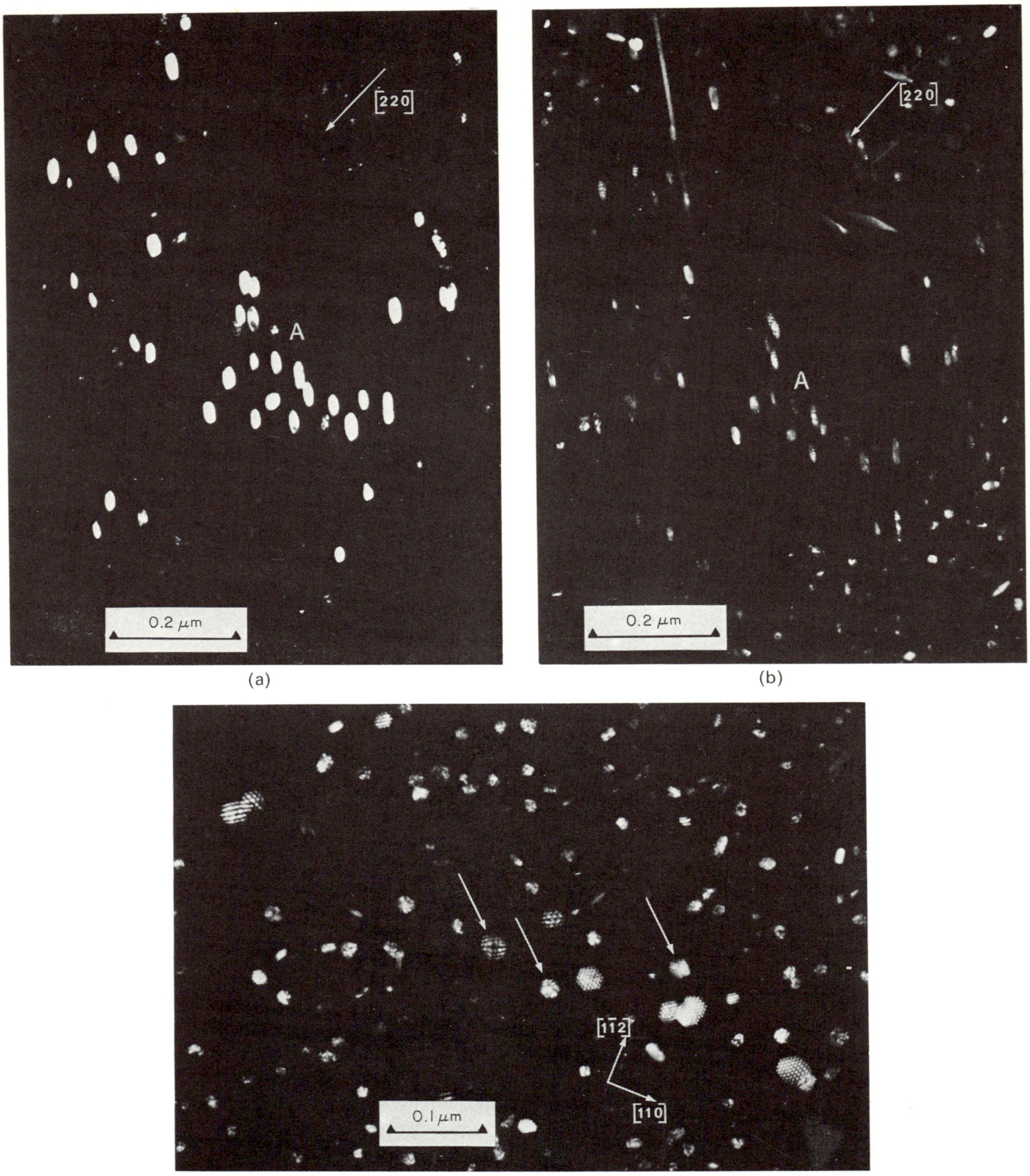

Figure 4.15 CDF images of the same specimen as figure 4.13(a), using the reflections circled in figure 4.13(b), (c). One variant of η' is visible in (a) together with η_2 particles showing Moiré patterns for CDF on reflection Y, figure 4.13(b) at A in (b). Shifts in the Moiré pattern (c) are interpreted as faults in η_2 [From Gjønnes and Simensen (1970), courtesy of *Acta Metallurgica*]

the $[1\bar{1}\bar{2}]$ pole is moved to the edge, and the orientation relationship

$$(0001)_{\eta'}//(1\bar{1}1)_{Al},\ (10\bar{1}0)_{\eta'}//(110)_{Al}$$

is obtained. Using $\boldsymbol{B}$ $[1\bar{1}\bar{2}]$ as before, the great circle containing all the η' reflecting planes is drawn in figure 4.16(a) and labelled $(1\bar{1}1)$ variant. Figure 4.17(a) shows the positions of the reflections to be expected. There are two other variants of this orientation relationship within $(0001)_{\eta'}//(1\bar{1}1)_{Al}$, either $[110]_{Al}//[01\bar{1}0]_{\eta'}$ or $[110]_{Al}//[\bar{1}100]_{\eta'}$, but these would produce patterns identical with that in figure 4.17(a) with different indexing.

4.9.2.2 *The* $(11\bar{1})$ *variant*

If the $(11\bar{1})$ pole is rotated to the centre of the stereogram, the $[1\bar{1}\bar{2}]$ pole is moved towards the edge and the orientation relationship

$$(0001)_{\eta'}//(11\bar{1})_{\text{Al}},\ (01\bar{1}0)_{\eta'}//(101)_{\text{Al}}$$

is obtained. The great circle, pole $[1\bar{1}\bar{2}]$ is shown in figure 4.16(a) and labelled $(11\bar{1})$ variant, and the precipitate reflections shown in figure 4.17(b) are generated. If all variants are allowed for the complete diffraction pattern shown in figure 4.17(c) is obtained which also includes the information in figures 4.16(c), 4.17(a). For simplicity, because most spots have several indices, this pattern has not been indexed. Of course, not all of these spots are necessarily visible for the following reasons.

(1) The intensity of some reflections may be comparable with the background because of structure factor effects.

(2) The curvature of the Ewald sphere combined with inexact ***B*** in the precipitate due to bending of the foil may reduce the effective intensity of some of the spots.

However, all of the spots shown in the computed diagram, figure 4.14(b), are included in figure 4.17(c).

This exercise may be repeated for the η_1 precipitate for the orientation relationship

$$(0001)_{\eta_1}//(110)_{\text{Al}},\ (10\bar{1}0)_{\eta_1}//(001)_{\text{Al}}$$

Superposition of the $(001)_{\eta_1}$ and $(110)_{\text{Al}}$ stereograms in figure 4.18(a),(b) for this orientation relationship shows that the nearest η_1 zone to the $[\bar{1}1\bar{2}]_{\text{Al}}$ zone is $[\bar{2}110]$ which is $\sim 6°$ rotation about

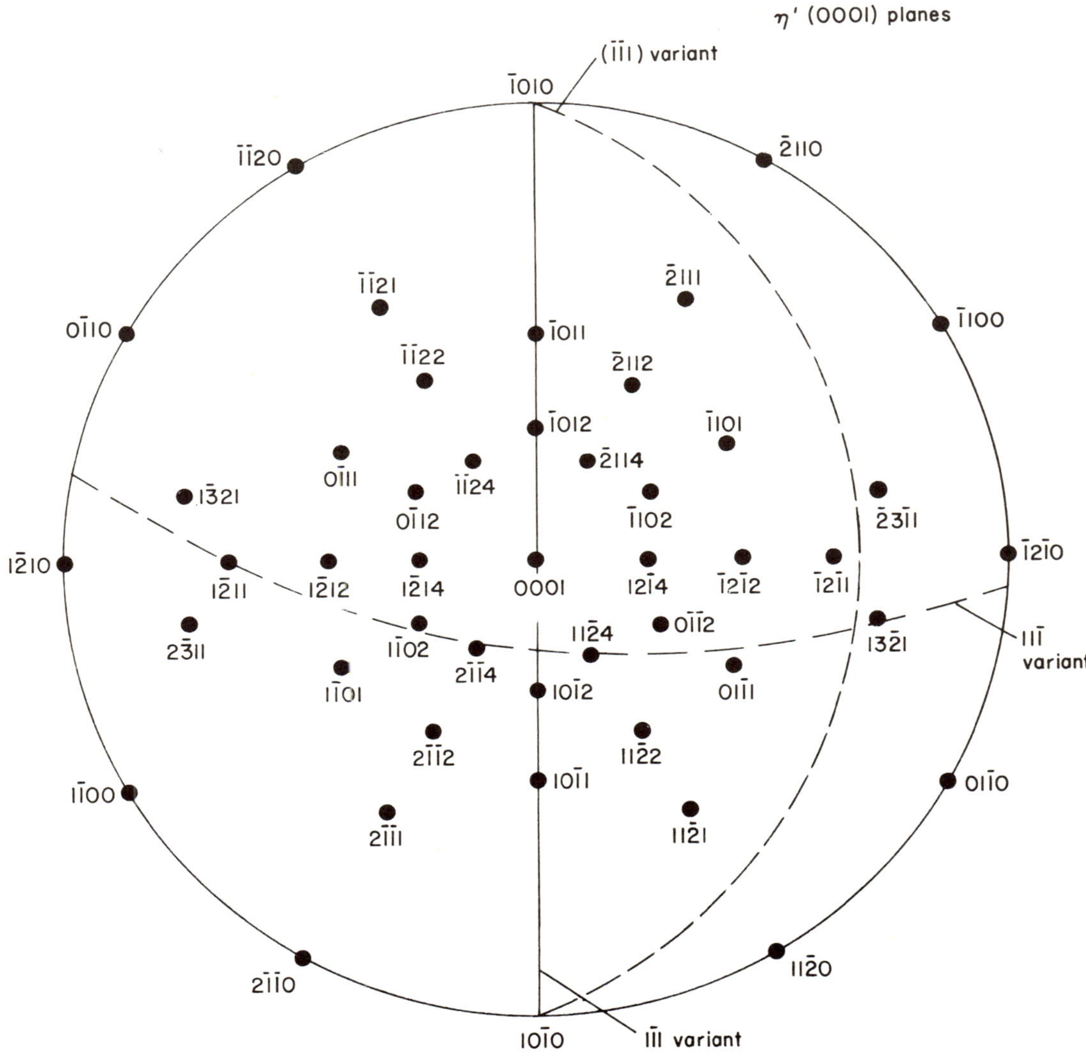

(a)

Figure 4.16 (a) and (b) are stereographic projections which, if superimposed, describe the η' orientation relationship in table 4.12. (c) The superimposed diffraction patterns for the $(0001)_{\eta'}//(1\bar{1}\bar{1})_{\text{matrix}}$ variant [Courtesy of A. J. de Ardo, Jr.]

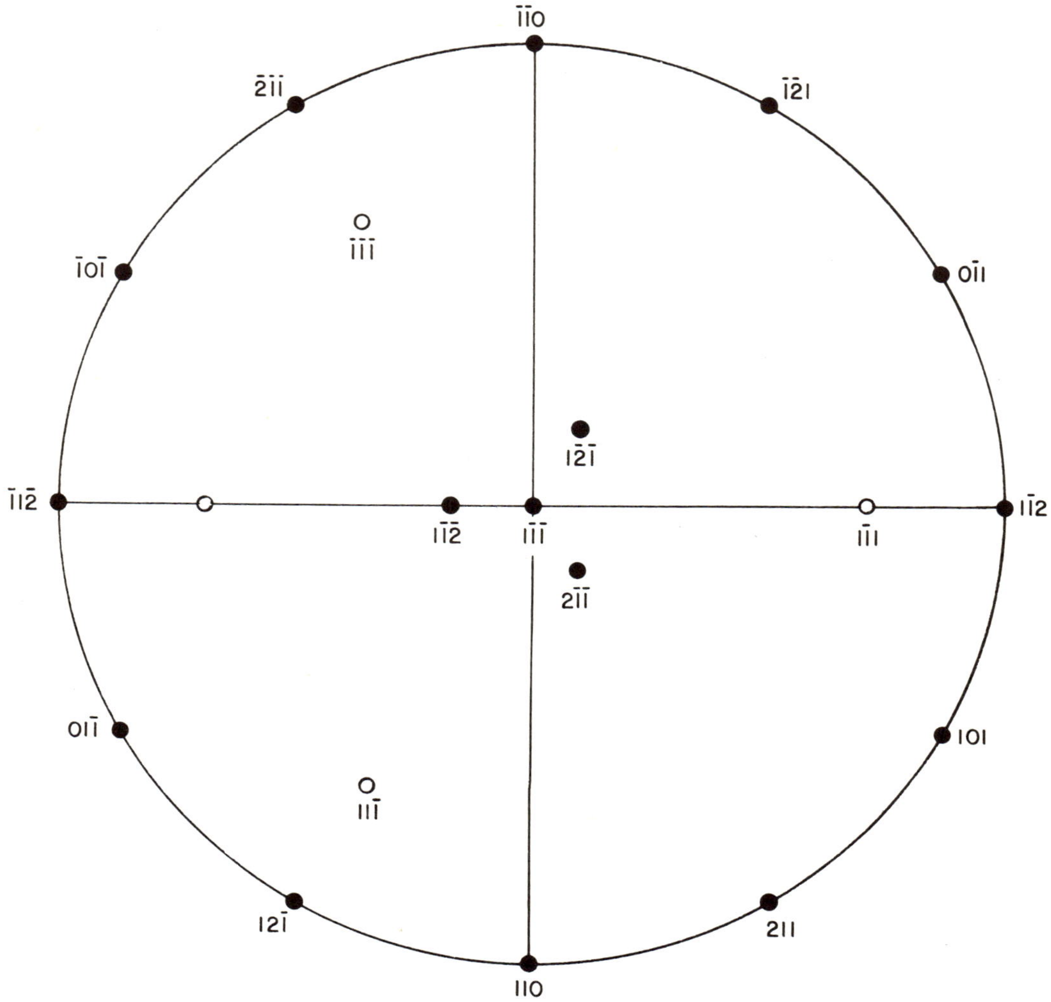

(b)

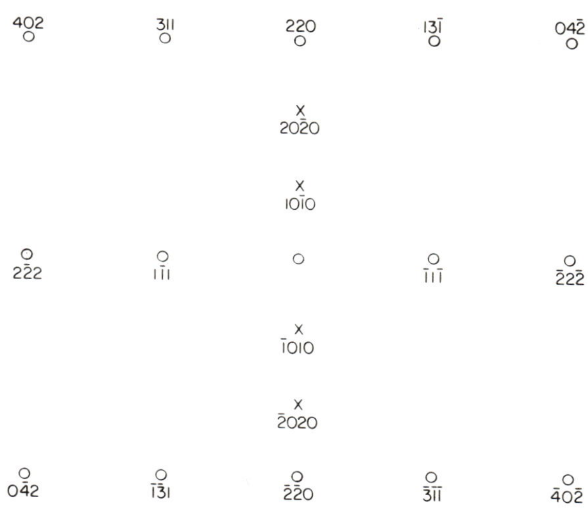

(c)

Figure 4.16 (*continued*)

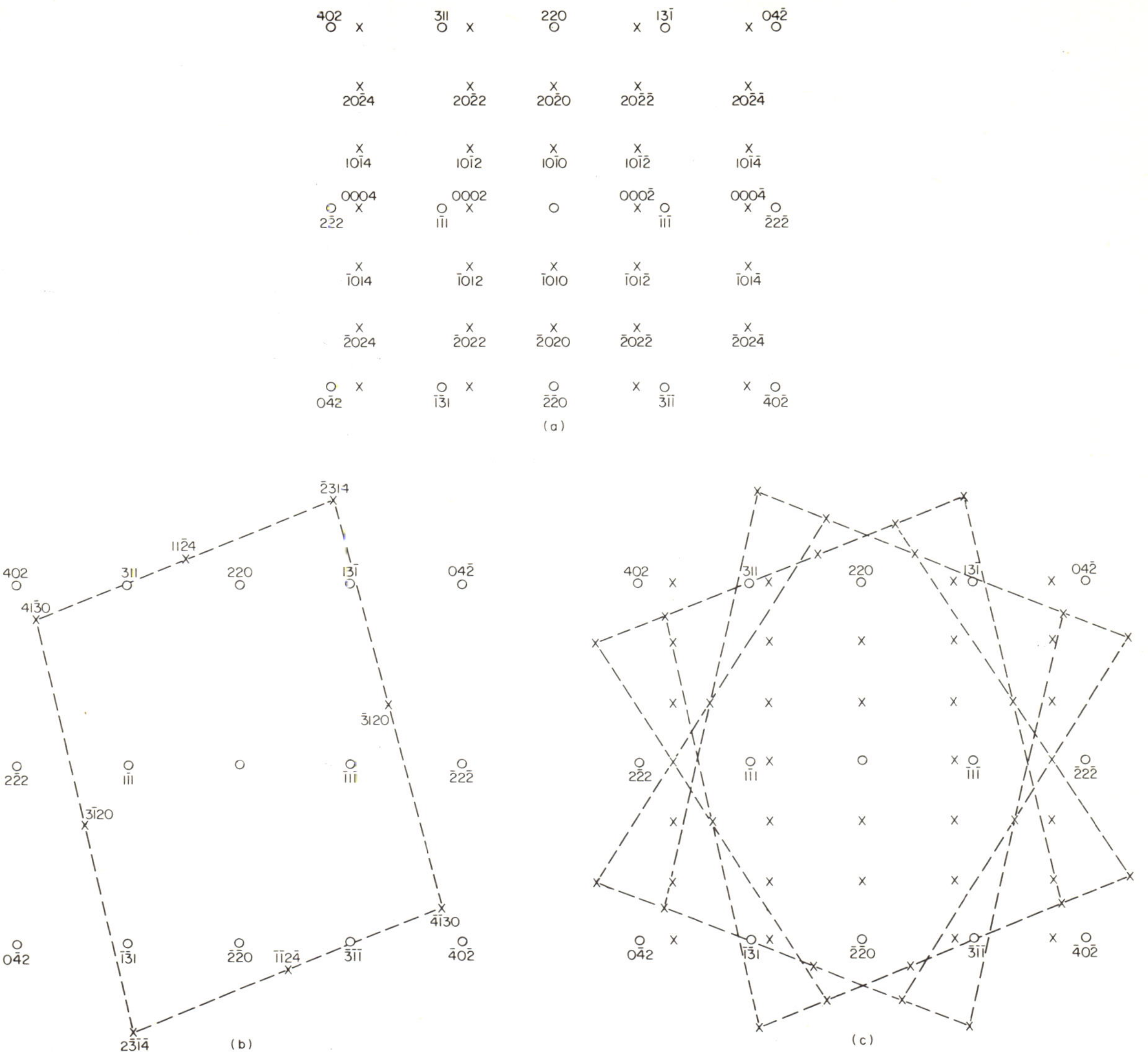

Figure 4.17 The superimposed diffraction patterns for (a) the $(0001)_{\eta'}//(1\bar{1}1)_{\text{matrix}}$ variant, (b) the $(0001)_{\eta'}//(11\bar{1})_{\text{matrix}}$ variant, (c) all variants [Courtesy of A. J. de Ardo, Jr.]

$[0001]_{\eta_1}$ from $[\bar{1}1\bar{2}]_{\text{Al}}$. Because of this rotation, only the row of $(0002)_{\eta_1}$ reflections would be expected to occur as shown in figure 4.18(c). When the other variants are taken into account, a very large number of spots occur in the $[\bar{1}1\bar{2}]_{\text{Al}}$ SADP. The filled-in squares in figure 4.18(d) show the more important spots which are known to occur in practice and which are present in the computed and experimental patterns, figures 4.14(b) and 4.13(c) respectively.

Clearly, the geometrical approach whereby the diffraction patterns are generated from the stereogram is time consuming and inaccurate because of accumulated errors arising from the stereographic manipulations, see appendix 1. These errors are particularly important when attempting to decide whether particular reflections should or should not be present for a particular ***B***.

4.9.3 *Quantitative Metallography of θ″ and θ′ Particles in an Al–4 wt% Cu Alloy (after J. D. Boyd)*

Both θ' and θ'' particles occur as thin discs lying on {100} planes of the aluminium matrix, as shown schematically in figure 4.19(a). BF images are unsatisfactory for measurement of particle dimensions because strain field contrast obscures the detail. Consequently, measurements must be made using CDF images obtained using precipitate reflections, see figure 4.19(b),(c).

The following measurements may be made.

(1) Mean particle diameter $\bar{D}$.
(2) Volume fraction $\bar{V}$.
(3) Interfacial area per unit volume A_V.
(4) Mean aspect ratio of the particles $\bar{K}$.

However, it is important to realise that, for a given CDF image, only a fraction of the particles may be visible, that is those contributing to the spot(s) within the objective aperture. Consequently, in figure 4.19(b),(c), one variant is clearly visible while a second is faintly visible because two precipitate reflections have been included in the objective aperture. Thus $\frac{2}{3}$ of the particles are visible in this case.

Experimentally, particularly for θ'' particles, the CDF technique is best employed with $\boldsymbol{B}\ \langle 100 \rangle$, that is the geometry in figure 4.19(a) with two sets of particles parallel to the electron beam and one further set perpendicular to it. This approach has the following advantages.

(1) Those particles parallel to the electron beam always exhibit the clearest images.

(2) Both the mean particle diameter and the mean aspect ratio may be measured directly from the same photographs provided magnification is accurately known.

Using the above approach a large number of values of both D and thickness are measured, for example 60–100, and a typical set of results are given in the histogram of figure 4.19(d). Assuming a normal distribution of errors the standard error of $\bar{K}$ is $\sigma/(N)^{1/2}$, where σ is the standard deviation of the results of N measurements. The error in $\bar{K}$ is $\pm 3\%$.

Determinations of $\bar{V}$ and A_V present considerable problems. Hilliard (1962) has shown that for disc-shaped particles

$$A_V = \frac{3\pi}{2} N_L - 2t\bar{D}N_V \tag{4.1}$$

$$\bar{V} = \frac{9\pi}{8}\left(L_1 - \frac{\pi}{2K} tN_L + \frac{2}{3\bar{K}} t^2 \bar{D} N_V\right) \tag{4.2}$$

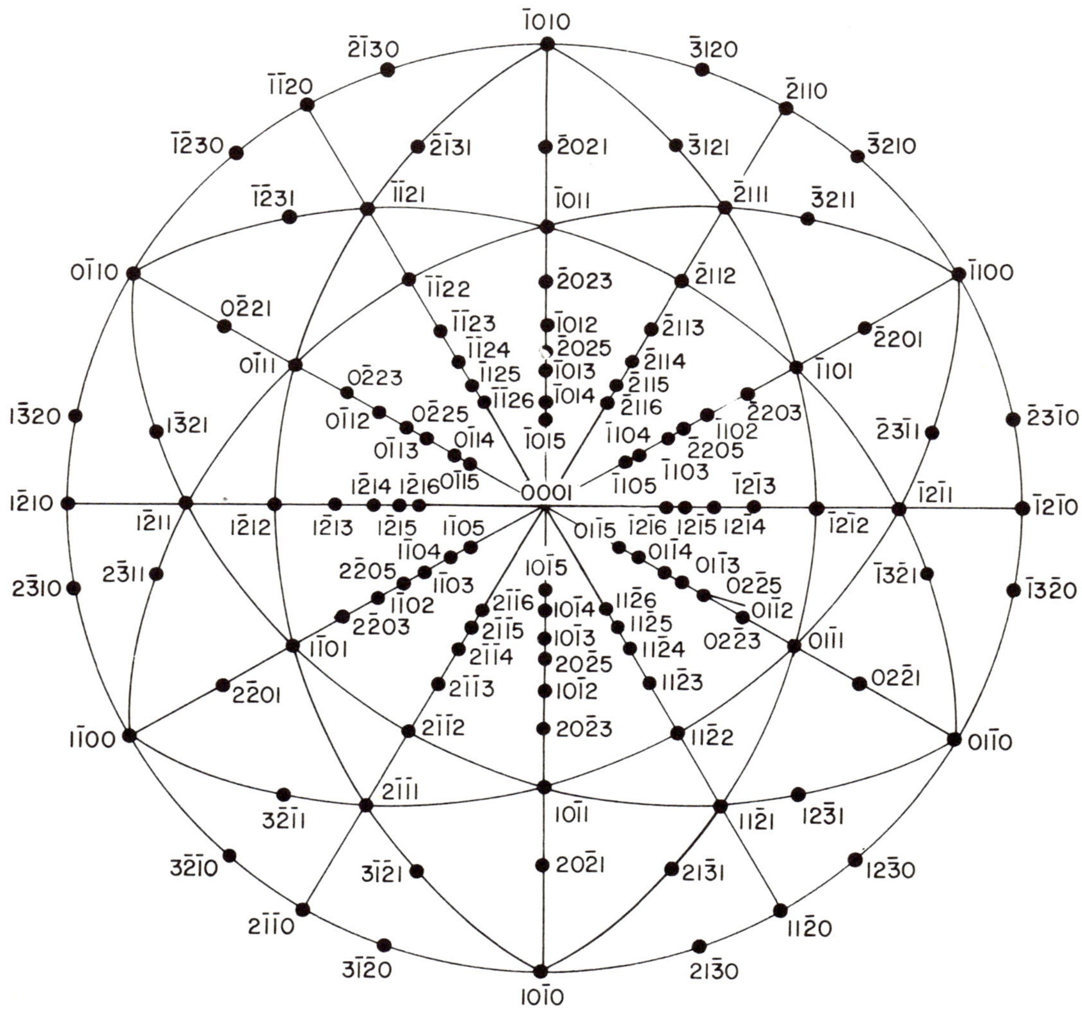

(a)

Figure 4.18 (a) and (b) are stereograms which, if superimposed, describe the η_1 orientation relationship in table 4.12; (c) the superimposed diffraction patterns for the $(0001)_{\eta_1}//(110)_{Al}$ variant; (d) all variants [Courtesy of A. J. de Ardo, Jr.]

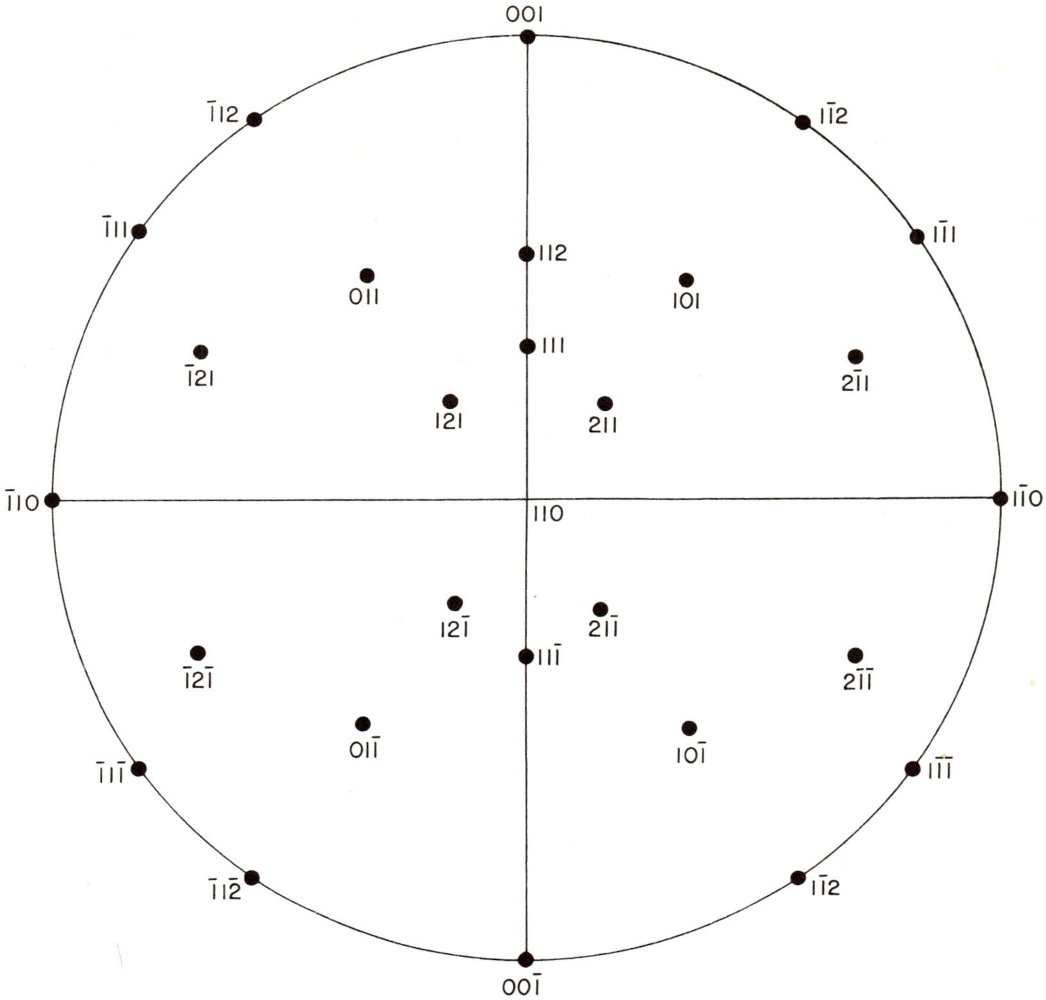

(b)

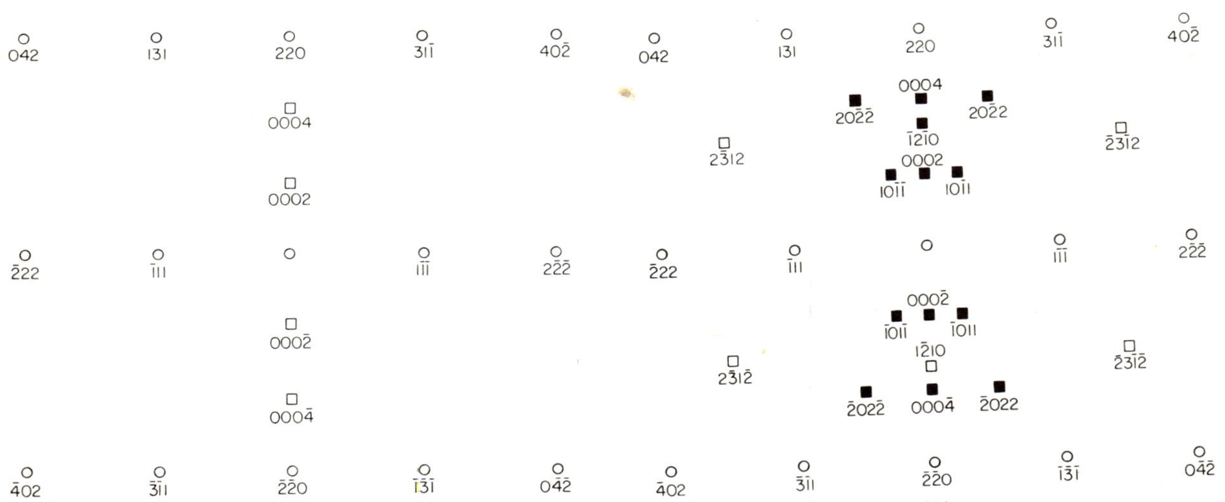

(c) (d)

Figure 4.18 (*continued*)

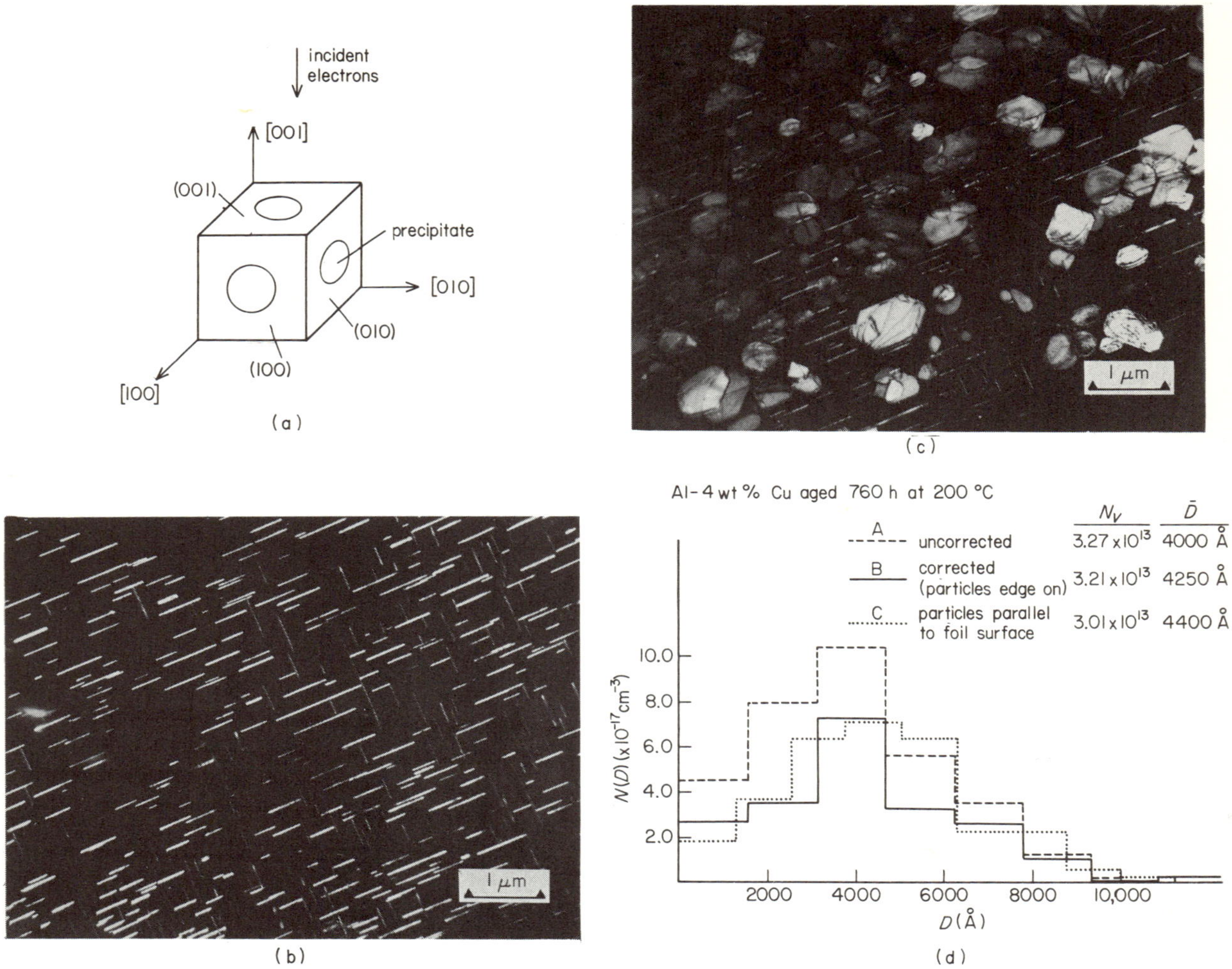

Figure 4.19 (a) The orientation of θ' precipitates in a thin foil of aged Al–Cu $\boldsymbol{B}[001]$. (b), (c) CDF images of an Al–4 wt% Cu alloy aged for 760 hours at 200 °C, obtained using $(10\bar{1})_{\theta'}$ and $(110)_{\theta'}$ reflections respectively. (d) A typical particle size distribution for θ', illustrating the effect of the correction procedure [Courtesy of J. D. Boyd]

where N_L is the number of intersections with an interface per unit length of random line; N_V, the number of particles per unit volume $= (t+\bar{c})/N_A$; L_1, the linear fraction of a random line that falls on a precipitate; t, the foil thickness; N_A, the number of particles per unit area of micrograph.

However, if this approach is used in the present case, there are significant difficulties.

(1) It is only valid if $t < \bar{D}$.

(2) Since the values of $\bar{c}$ and N_V are obtained by plotting N_A versus t it is difficult to obtain specimens with a large enough range of thickness to compensate for errors in t (which are $\pm 15\%$).

(3) Rapid variations in thickness can easily occur over quite short distances, that is ~ 1 μm.

Values for $\bar{V}$ and A_V may also be obtained directly from the measured particle size distributions $N(d)$—using the relations

$$N_V = \sum_D N(D)\Delta D \qquad (4.3)$$

$$\bar{D} = \frac{1}{N_V}\sum_D \frac{N(D)\Delta D}{D} \qquad (4.4)$$

$$A_V = \left(\frac{1}{2}+\frac{l}{K}\right)\sum_D D^2 N(D)\Delta D \qquad (4.5)$$

$$\bar{V} = \frac{\pi}{4K}\sum_D D^3\, N(D)\Delta D \qquad (4.6)$$

where $N(D)\Delta D$ is the actual number of particles per unit volume with diameters in the region $D \pm \Delta D/2$. The experimentally determined values for $N(d)$ are biassed towards small diameters because a fraction of the particles cut the surface and must therefore be measured at less than their correct diameter. The relationships between $N(d)$ and the true size distribution is given by the equation

$$\{N(D)\Delta D\}_{D=d} = \frac{N(d)\Delta d - \mu(d)\Delta d}{t} \qquad (4.7)$$

where $\mu(d)$ is the contribution from particles centred

outside the foil. The real particle size distribution $N(D)$ may be calculated from the measured size distribution using the method derived by Scheil (1935). Thus the measured size distribution may be divided up into intervals of width Δd and it is assumed that particles with the maximum diameter can be accurately measured, that is their centres are in the foil. Therefore, the real and measured particle sizes coincide at the upper end of the size distribution, that is

$$\{N(D)\Delta D\}_{D=d_{max}} = \left\{\frac{N(d)\Delta d}{t}\right\}_{d=d_{max}} \tag{4.8}$$

For the precipitates in Al–Cu, $t/(D)^{1/2} \approx 0.5-2$ and overlap effects (edge-on particles) could be neglected. As a result we may assume the relation (Scheil, 1935)

$$\mu(d) = \sum_{D=d}^{\infty} d(D^2-d^2)^{-1/2}N(D)\Delta D \tag{4.9}$$

Hence the number of particles in each interval that arise from intersections with particle of diameter $d = d_{max}$ was calculated and subtracted from the measured particle sizes $N(d)$. The number of particles per unit volume in the second-largest interval could then be calculated and the procedure repeated until the complete actual size distribution $N(D)$ was determined. These calculations may be easily computerised.

The values of $N(d)$ were determined by measuring one set of particles visible under one reflection, for example figure 4.19(b), and multiplying the result by three to account for all $\{100\}$ habit planes. Figure 4.19(d) shows the histogram of the size measurements on one set of edge-on particles in figure 4.19(b). The dashed line is uncorrected $N(d)$ and the solid line using the Scheil correction described above which reduces the number of small-diameter particles by $\sim 30\%$.

The accuracy of this correction procedure may be checked by measuring particles lying parallel to the foil surface like those in figure 4.19(c). Because the particle thickness is so small, very few particles intersected the surface and the correction need not be applied. The dotted line in figure 4.19(d) shows the size distribution so obtained and it is close to the corrected line for edge-on particles (solid line). The discrepancy in the two methods is 15% in N_V and 3% in $\bar{D}$. In this case, $\bar{D} \sim t$ for $\bar{D} < t$ the accuracy would be greater because of the smaller correction.

The calculation of $\bar{D}$, N_V, A_V and $\bar{V}$ can best be illustrated by considering a specific example. The dispersion of θ'' precipitates shown in figure 4.20 was measured to determine $N(d)$. The foil thickness is 1000 Å and $\bar{K}$ was 20. With the aid of a computer, the quantities $N(D)\Delta D$ and their related terms for the above equations were calculated, see table 4.13.

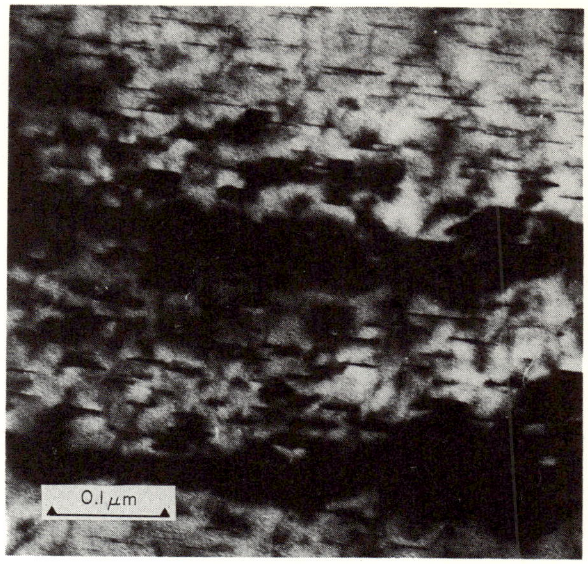

Figure 4.20 A BF image of Al–4 wt % Cu aged for 10 days at 130 °C followed by 32 hours at 157 °C

Table 4.13 Calculation of dispersion parameters

D (Å)	$N(d)\Delta d$ (cm$^{-2}\times 10^{-10}$)	$N(D)\Delta D$ (cm$^{-3}\times 10^{-15}$)
23.2	3.32	2.95
69.5	7.43	6.56
116	8.22	7.02
162	9.37	7.99
209	12.09	11.75
255	10.05	9.72
301	7.08	6.64
347	3.58	3.32
393	2.01	1.87
440	1.40	1.37
486	0.26	0.26

Al–4 wt % Cu alloy solution treated 6 hours at 525 °C, aged 10 days at 130 °C then 32 hours at 157 °C.

$\sum_D DN(D)\Delta D = 1.22\times 10^{11}\ \text{cm}^{-2}$

$\sum_D D^2N(D)\Delta D = 2.10\times 10^{3}\ \text{cm}^{-1}$

$\sum_D D^3N(D)\Delta D = 0.898$

$\sum_D N(D)\Delta D = 5.95\times 10^{16}\ \text{cm}^{-3}$

The same measurements were made on other micrographs until a sufficient number of particles had been measured to make the sample errors in $N(d)$ negligible, ~ 500 particles. The final calculated values for $\bar{D}$, N_V and $\bar{V}$ are listed in table 4.14. The error in $N(D)$ was principally due to the error of 14–20% in t and the possible error in N_V was estimated to be $\sim 15\%$. Since the error in $D \ll N_V$ the values of $\bar{D}$, A_V and $\bar{V}$ are expected to be accurate to within $\sim 15\%$.

Table 4.14 Alloy treatment as for table 4.13

Parameter	ξ_9	Value	Estimated error
K		20	± 0.6
N_V	3	$5.95 \times 10^{16}\ \mathrm{cm}^{-3}$	$\pm 0.9 \times 10^{16}\ \mathrm{cm}^{-3}$
$\bar{D}$	4	205 Å	± 31 Å
A_V	5	$5.3 \times 10^5\ \mathrm{cm}^{-1}$	$\pm 0.8 \times 10^5\ \mathrm{cm}^{-1}$
$\bar{V}$	6	0.035	$\pm 0.005\ \mathrm{cm}^{-1}$
A_V	1	$2.7 \times 10^5\ \mathrm{cm}^{-1}$	$\pm 5.7 \times 10^5\ \mathrm{cm}^{-1}$
$\bar{V}$	2	0.17	$\pm 0.34 \times 10^5\ \mathrm{cm}^{-1}$

Values of $\bar{V}$ and A_V measured by the other approach using equations (4.1) and (4.2) are also included in table 4.14. The difference is large but is within the error of the latter measurement.

Note that for large values of $\bar{D}$ this latter method will have larger errors. Nevertheless, within the limit $\bar{D} < t$ it is reasonably accurate for this type of situation.

4.10 Cobalt Alloys

These alloys are used commercially for both structural and magnetic materials. In general, structural materials are strengthened by carbide precipitates whereas the magnetic materials such as Cunico and Vicalloy contain a high density of particles, thought to be formed by spinodal decomposition (Tufton, 1963). In this section the spinodal wavelength in a Co–Ti alloy is measured.

4.10.1 *The Measurement of the Composition Modulation Wavelength in an Aged Co–8 wt% Ti Alloy (after M. N. Thompson)*

As pointed out in section 2.17.2.1 spinodal decomposition produces satellite spots in the SADP symmetrically displaced about normal $\{hkl\}$ reflections. Figure 4.21 shows a typical case for a 8 wt% Ti Alloy quenched from 1200 °C then aged at 875 °C for 10 minutes and quenched to room temperature. The SADP is shown in figure 4.21(b) together with a magnified picture of the satellites at the $(0\bar{2}\bar{2})$ matrix reflection. Figure 4.21(a) shows the distribution of Co_3Ti particles in the specimen. This image was obtained using CDF on the $\{100\}$ Co_3Ti superlattice reflection position, although no spot could be detected on the screen. Using equation (2.33) the wavelength of the modulated structure may be obtained using values for $\boldsymbol{g}$ and $\Delta\boldsymbol{g}$ taken from a microdensitometer trace of the

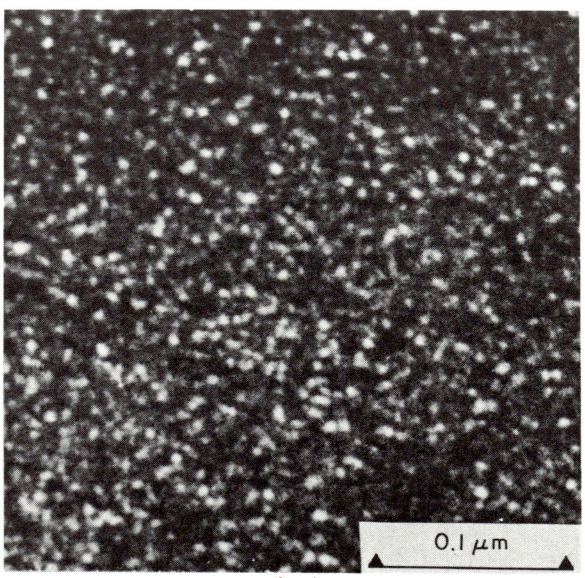

(a)

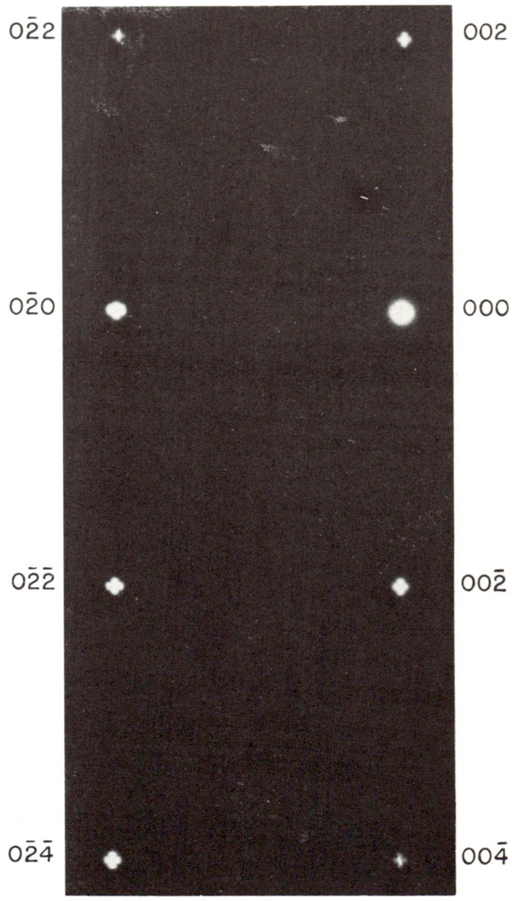

(b)

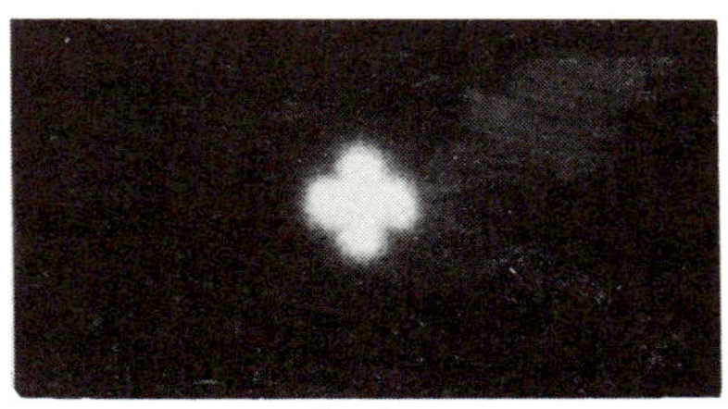
(c)

Figure 4.21 (a) A homogeneous distribution of coherent Co_3Ti precipitates produced during quenching a Co–8 wt% Ti specimen from 875 °C, CDF using Co_3Ti reflection. (b) The SADP from an area similar to (a) and (c) satellites about the $(0\bar{2}\bar{2})$ reflection original plate. Considerable care is necessary in alignment of the direction of travel of the microdensitometer. Maximum intensity on both $\pm \boldsymbol{g}$ for a given trace is necessary before the measurement can be taken. The value of λ for the trace in figure 4.22 is 62 ± 4 Å.

4.11 Nickel Alloys

Nickel-base alloys are used in a wide variety of applications, one of the most important being gas turbines. It is from this series of high-strength high-temperature alloys that most of the examples in this section are taken.

Nickel-base alloys designed for high-temperature service are almost without exception precipitation hardened. The major precipitate is the ordered γ' phase which has the $L1_2$ crystal structure and occurs as coherent particles (volume fraction ~40–60%). However, a number of carbide phases are also present, see table 4.15. In addition, a number of intermetallic phases are found during service such as sigma, mu and Laves phases. Finally some nickel-base alloys are dispersion hardened with inert particles. Thus TD nickel contains ThO_2 particles and has useful high-temperature strength.

A number of examples of quantitative analysis of these materials are given below. It should be noted that, although γ' particles in these alloys may be < 100 Å in diameter for maximum strength, quantitative analysis of dislocation Burgers' vectors, sense of stacking faults, etc., are usually performed for much coarser particles when it is easier to obtain consistent results.

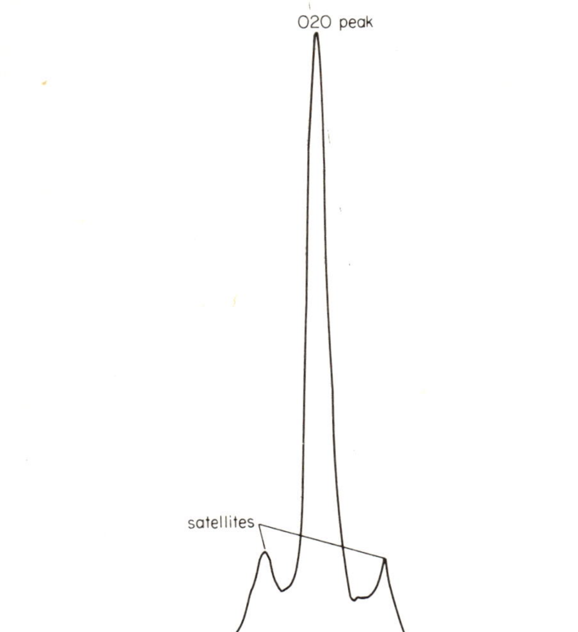

Figure 4.22 Shows a schematic diagram of a microdensitometer trace across the satellites in figure 4.21(c)

Table 4.15 Carbides/nitrides in Nimonic alloys

Nimonic alloy	Carbide	
80A	Cr_7C_3	grain boundaries
	$Cr_{23}C_6$	transformed Cr_7C_3
	TiC/N	inclusions
90	$Cr_{23}C_6$	grain boundaries
	TiC/N	inclusions
105 and 115	$Cr_{23}C_6$	grain boundaries
	M_6C	formed from $M_{23}C_6$ crystalline Ni, Mo, Cr
	TiC/N	inclusions

4.11.1 *Identification of Cr_7C_3 in Nimonic 75 (after K. S. Probert)*

This alloy consists of a Ni–20 wt % Cr solid solution containing carbon, to which a small amount of Ti is added. In effect, it is a solid solution strengthened matrix containing carbide particles present both at grain boundaries and within the grain. Figure 4.23(a) shows a typical extraction replica taken from material aged for 16 hours at 850 °C and figure 4.23(b) shows a transmission electron micrograph from the same material. Electron diffraction patterns taken from individual carbide particles in the replica for different $\boldsymbol{B}$ are shown in figure 4.24(a)–(c).

The structure of Cr_7C_3 is complex pseudo-hexagonal with $a = 13.980$ Å and $c = 4.523$ Å. The interplanar spacings of this phase containing some iron are shown in table 4.16. Within the experimental error it is possible to index the patterns in figure 4.24(a)–(c) using table 4.16 as shown in figure 4.24(d)–(f), using the method of ratios described in section 2.7.2.1.

Table 4.16 Interplanar spacings of Cr_7C_3 containing Fe

hkil	*d* (Å)	*hkil*	*d* (Å)	*hkil*	*d* (Å)	*hkil*	*d* (Å)
$31\bar{4}1$	2.696	$52\bar{7}1$	1.782	$8,2,\overline{10},0$	1.321	$7,4,\overline{11},1$	1.210
$32\bar{5}1$	2.367	$44\bar{8}0$	1.748	$32\bar{5}3$	1.325	$9,2,\overline{11},0$	1.193
$33\bar{6}0$	2.330	$32\bar{5}2$	1.754	$90\bar{9}1$	1.289	$8,3,\overline{11},1$	1.186
$42\bar{6}0$	2.288	$61\bar{7}1$	1.709	$54\bar{9}2$	1.279	$52\bar{7}3$	1.190
$41\bar{5}1$	2.281	$70\bar{7}1$	1.612	$50\bar{5}3$	1.280	$5,5,\overline{10},2$	1.189
$10\bar{1}2$	2.223	$33\bar{6}2$	1.623	$33\bar{6}3$	1.266	$10,0,\overline{10},1$	1.170
$11\bar{2}2$	2.152	$60\bar{6}2$	1.506	$5,6,\overline{11},0$	1.269		
$20\bar{2}2$	2.119	$71\bar{8}1$	1.511	$9,1,\overline{10},0$	1.269		
$50\bar{5}1$	2.135	$80\bar{8}1$	1.435	$8,2,\overline{10},1$	1.268		
$42\bar{6}1$	2.042	$61\bar{7}2$	1.430	$42\bar{6}3$	1.259		
$60\bar{6}0$	2.018	$21\bar{3}3$	1.432	$80\bar{8}2$	1.258		
$21\bar{3}2$	2.027	$62\bar{8}2$	1.348	$72\bar{9}2$	1.238		
$51\bar{6}5$	1.960	$40\bar{4}3$	1.350	$51\bar{6}3$	1.239		
$22\bar{4}2$	1.899	$81\bar{9}1$	1.352	$83\bar{9}0$	1.229		
$60\bar{6}1$	1.843	$90\bar{9}0$	1.345	$10,0,\overline{10},0$	1.211		
$40\bar{4}2$	1.812	$6,4,\overline{10},1$	1.328	$60\bar{6}3$	1.208		

Crystal structure is pseudo-hexagonal.

Unit cell: $a = 13.98$ Å, $c = 4.523$ Å. Recalculated by Dyson and Andrews (1969).

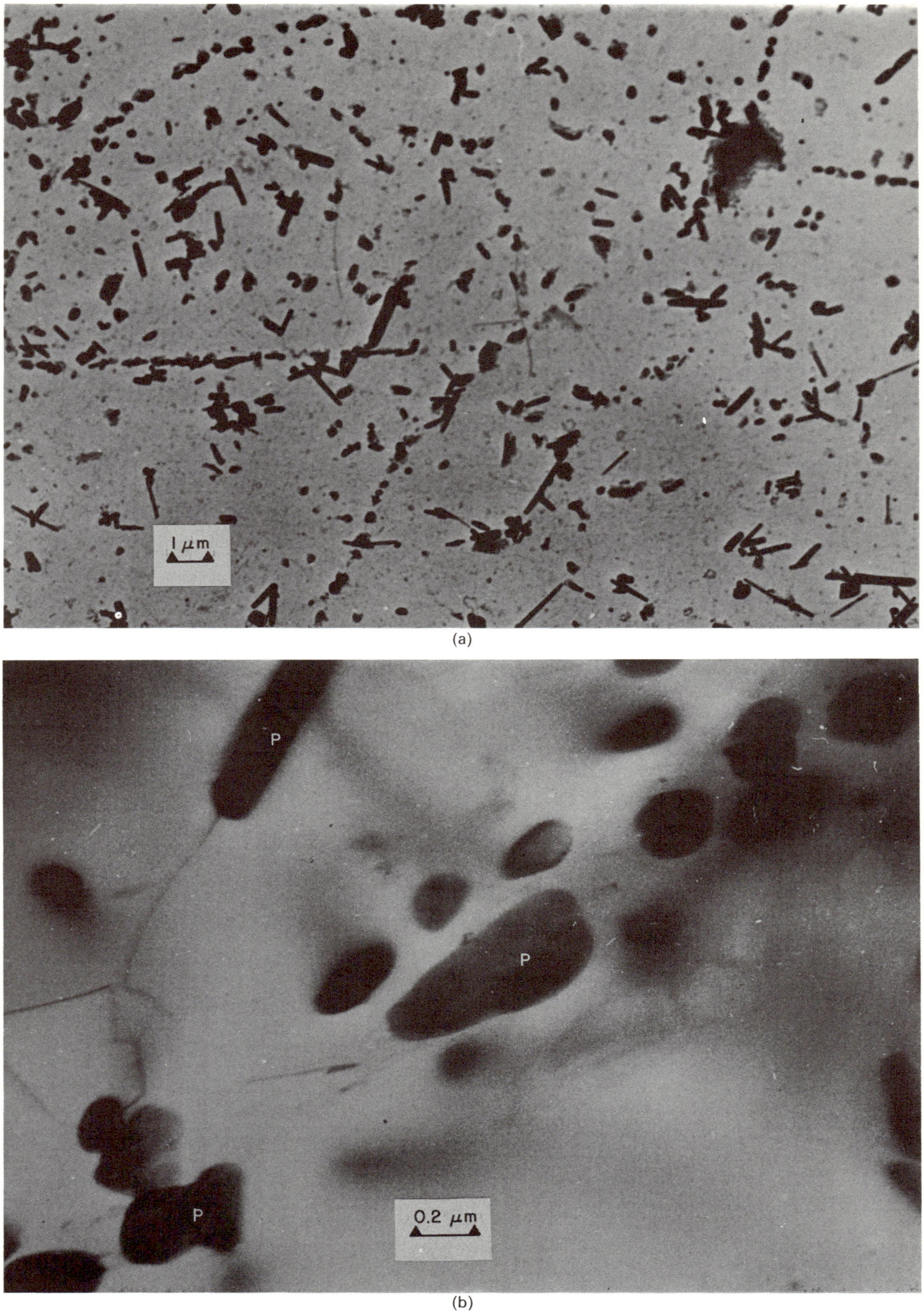

Figure 4.23 (a) An extraction replica and (b) a BF image taken from a Nimonic 75 specimen aged at 850 °C for 16 hours showing carbide particles P [Courtesy of K. S. Probert]

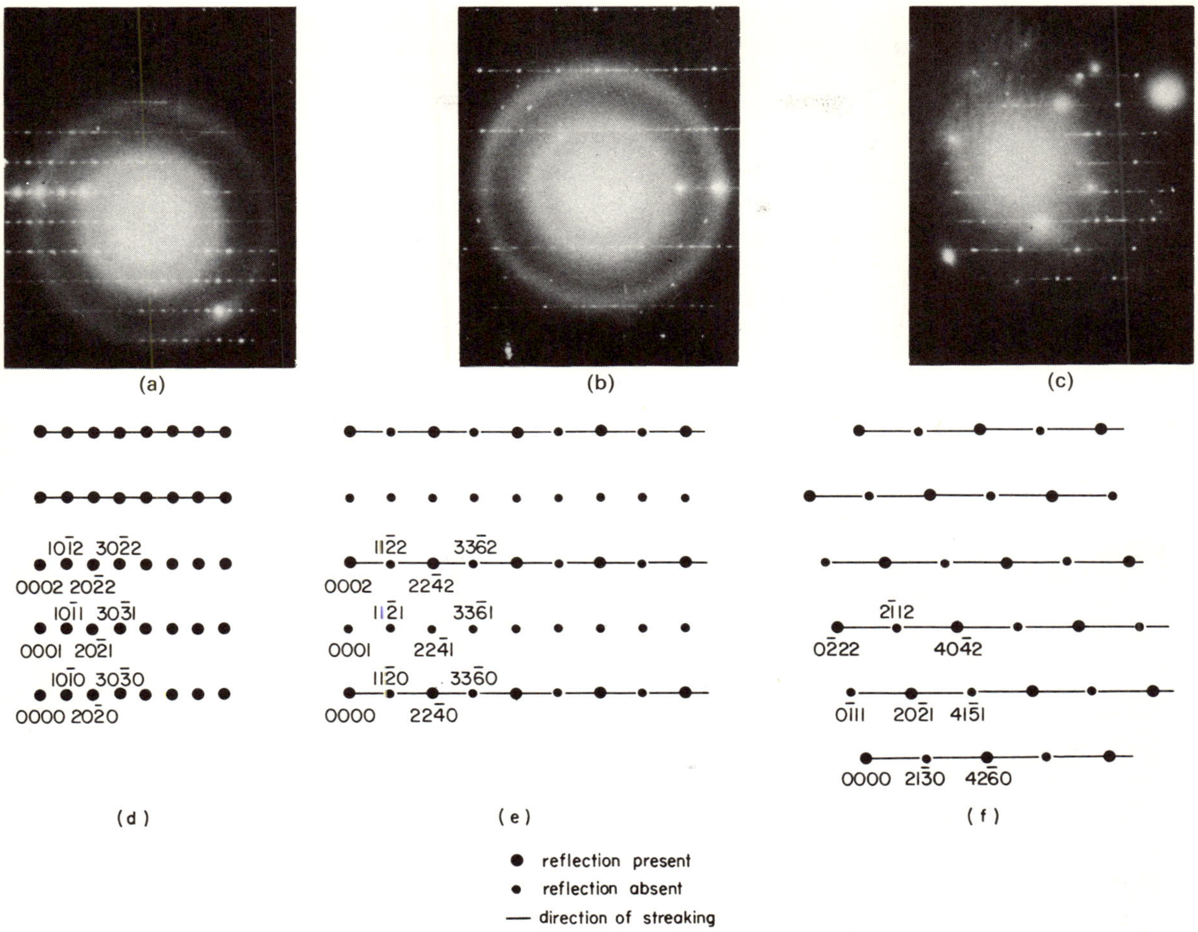

Figure 4.24 A selection (a)–(c) of diffraction patterns taken from carbide particles in quenched and aged Nimonic 75, (d)–(f) show the corresponding patterns indexed as Cr_7C_3 [Courtesy of K. S. Probert]

Many reflections are absent compared with those to be expected from a material with a primitive hexagonal crystal structure. However, in this, as in many other phases, the complexity of the crystal structure precludes calculation of structure factors so these absences cannot be checked.

The presence of streaks in the SADP lying in directions in the basal plane of the Cr_7C_3 crystal structure is interpreted in terms of stacking faults planes on and perpendicular to the basal plane by Dyson and Andrews (1969), that is $\{0001\}$ and $\{11\bar{2}0\}$. Beech and Warrington (1966) have used this characteristic streaking as a means of distinguishing this carbide from others.

4.11.2 *The Determination of the Burgers Vector of Partial Dislocations in MAR-M-200 (after B. H. Kear)*

As pointed out in section 3.4.5 application of conventional invisibility criteria to the determination of the direction of $\boldsymbol{b}_p$ for partial dislocations presents considerable difficulty. The rule $\boldsymbol{g} \cdot \boldsymbol{b}_p = 0$ or $\pm\frac{1}{3}$ is invisible is simple but the behaviour under $\boldsymbol{g} \cdot \boldsymbol{b}_p = \pm\frac{2}{3}$ is complex. As pointed out in section 3.4.5 if computer matching is not to be used, the most satisfactory method of determining the direction of $\boldsymbol{b}_p$ involves obtaining several cases of invisibility which can be all ascribed to $\boldsymbol{g} \cdot \boldsymbol{b}_p = 0, \pm\frac{1}{3}$ together with several cases of visibility for $\boldsymbol{g} \cdot \boldsymbol{b}_p =$ integer. An example of the latter approach in determining the direction of $\boldsymbol{b}_p$ is shown in this section for partial dislocations in a nickel-base alloy. Using the following approximate rules the magnitude of $\boldsymbol{b}_p$ was also determined.

Partial dislocations are as follows.

(1) Invisible if $\boldsymbol{g} \cdot \boldsymbol{b}_p = -\frac{2}{3}$ but visible if $\boldsymbol{g} \cdot \boldsymbol{b}_p = +\frac{2}{3}$ provided $w \gtrsim 0.7$ (Silcock and Tunstall, 1964).

(2) Invisible $\boldsymbol{g} \cdot \boldsymbol{b}_p = +\frac{4}{3}$ but visible if $\boldsymbol{g} \cdot \boldsymbol{b}_p = -\frac{4}{3}$ provided $w \gtrsim 0.7$ (Oblak and Kear, 1972).

As pointed out in section 3.4.5, these rules are not absolute because cases of anomalous contrast behaviour occur for $\boldsymbol{g} \cdot \boldsymbol{b}_p = \pm\frac{2}{3}$. However, in this example a cross-check on $\boldsymbol{b}_p$ is obtained from the geometry of the dislocation arrays and the direction of $\boldsymbol{b}_p$ is obtained in any case using only $\boldsymbol{g} \cdot \boldsymbol{b}_p = 0$ or 1 situations.

The example described here is that due to Kear *et al.* (1969) and describes the identification of intrinsic/extrinsic stacking fault pairs with bounding partial dislocations, net **b** ⟨112⟩. Figure 4.25 shows a section parallel to the critical slip plane (1$\bar{1}$1) in a single crystal of MAR-M-200 strained ~1.4% in primary creep at 750 °C. The dark areas were shown to be stacking faults using the fringe images observed in foils taken perpendicular to the tensile axis (Kear *et al.*, 1969). These stacking faults will be shown to be bounded by partial dislocations which are labelled 1–4 in figure 4.25(a). The light region between dislocations 2 and 3, figure 4.25(a), is not faulted. In some cases, for example at B, the stacking fault is contained within the γ' particles which in this case are approximately spherical. Dislocation 4 tends to lie in the particle–matrix interface and so, to a lesser extent, do dislocations 2 and 3. Knowledge of the foil orientation in relation to the applied stress enabled dislocations 1 to be identified as the leading dislocation in the group.

The orientation of the foil and the indices of the operative reflections were uniquely determined (Kear *et al.*, 1969) using the following.

(1) The projected width of the stacking faults on ($\bar{1}$11) the primary slip plane and (1$\bar{1}$1) the critical slip plane and (111) or ($\bar{1}\bar{1}$1).

(2) The sense of inclination of the faults knowing the top and bottom of the foil by comparison of BF and DF images (section 3.6.1.1). Unique

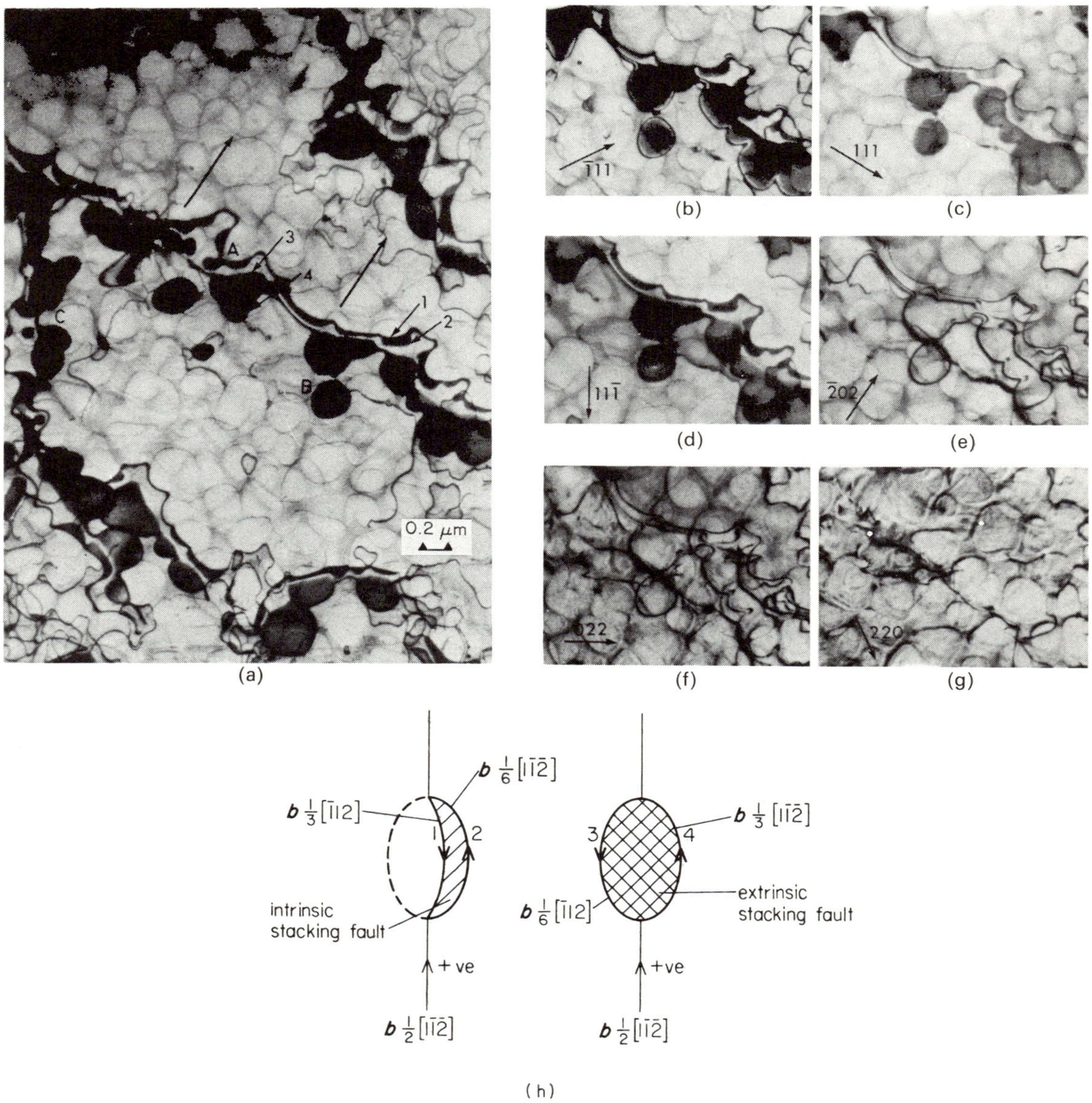

Figure 4.25 Creep-induced intrinsic/extrinsic fault pairs in a precipitation hardened single-crystal nickel-base alloy, sectioned parallel to the (1$\bar{1}$1) critical plane. The reflections used are indicated. (h) A schematic diagram describing the intersection of dislocations with γ' particles [From Kear *et al.* (1969), courtesy of *Quarterly Transactions of the American Society for Metals*]

indexing of this type enabled the sign of $\boldsymbol{g} \cdot \boldsymbol{b}_p$ to be used to determine not only the direction but also the magnitude of $\boldsymbol{b}_p$ for the partial dislocations 1–4.

First it is necessary to ascertain whether the dislocations are Frank partials, $\boldsymbol{b}_p \pm \frac{1}{3}[1\bar{1}1]$ or Shockey partials, glissile in $(1\bar{1}1)$, that is $\boldsymbol{b}_p = \pm\frac{1}{6}[\bar{1}21]$, $\pm\frac{1}{6}[\bar{2}\bar{1}1]$ or $\pm\frac{1}{6}[\bar{1}12]$. If the dislocations were Frank partials, $\boldsymbol{b}_p \pm \frac{1}{3}[1\bar{1}1]$, $\boldsymbol{g} \cdot \boldsymbol{b}_p = 0$ for $(\bar{2}02)$, (022) and (220) reflections, that is the dislocations should be invisible or show double-dotted images all three reflections, but this is not so, see figures 4.25(e),(g),(f). Thus the dislocations must be Shockley partials and the direction of $\boldsymbol{b}_p$ may be determined using the complete invisibility under the (220) reflection. In this case, $\boldsymbol{g} \cdot \boldsymbol{b}_p = \pm 1$ for $\boldsymbol{b}_p = \pm\frac{1}{6}[\bar{2}\bar{1}1]$ and $\pm\frac{1}{6}[\bar{1}21]$ but $\boldsymbol{g} \cdot \boldsymbol{b}_p = 0$ for $\boldsymbol{b}_p \pm \frac{1}{6}[\bar{1}12]$. Thus all four dislocations have $\boldsymbol{b}_p$ in the same direction. The magnitude of $\boldsymbol{b}_p$, that is $\pm\frac{1}{6}[\bar{1}12]$ or $\pm\frac{1}{3}[\bar{1}12]$ may be determined by consideration of the image behaviour under $(\bar{1}11)$, (111) and $(11\bar{1})$ reflections, with the aid of table 4.17.

Table 4.17 Values of $\boldsymbol{g} \cdot \boldsymbol{b}_p$ for $\boldsymbol{b}_p \pm \frac{1}{6}[\bar{1}12]$ and $\pm\frac{1}{3}[\bar{1}12]$

$\boldsymbol{g}$	$\boldsymbol{b}_p \pm \frac{1}{6}[\bar{1}12]$ $\boldsymbol{g} \cdot \boldsymbol{b}_p$	$\boldsymbol{b}_p \pm \frac{1}{3}[\bar{1}12]$ $\boldsymbol{g} \cdot \boldsymbol{b}_p$
$\pm[\bar{1}11]$	$\pm\frac{2}{3}$	$\pm\frac{4}{3}$
$\pm[11\bar{1}]$	$\pm\frac{1}{3}$	$\pm\frac{2}{3}$
$\pm[111]$	$\pm\frac{1}{3}$	$\pm\frac{2}{3}$

In comparing the behaviour of dislocations 2 and 3 under $(\bar{1}11)$, $(11\bar{1})$ and (111) reflections we see that they are both invisible for (111) and $(11\bar{1})$ reflections, that is $\boldsymbol{g} \cdot \boldsymbol{b}_p = \pm\frac{1}{3}$ and $\boldsymbol{b}_p = \pm\frac{1}{6}[\bar{1}12]$. However, dislocation 2 is invisible for the $(\bar{1}11)$ reflection ($\boldsymbol{g} \cdot \boldsymbol{b}_p = -\frac{2}{3}$) while dislocation 3 is visible ($\boldsymbol{g} \cdot \boldsymbol{b}_p = +\frac{2}{3}$). Consequently, $\boldsymbol{b}_p = \frac{1}{6}[1\bar{1}\bar{2}]$ for dislocation 2 where $\boldsymbol{b}_p = \frac{1}{6}[\bar{1}12]$ for dislocation 3. The contrast behaviour for dislocations 1 and 4 is shown in table 4.18 for the three $\{\bar{1}11\}$ reflections.

Table 4.18 Visibility of dislocations 1 and 4, figure 4.25

$\boldsymbol{g}$	Dislocation 1	Dislocation 4	Dislocation 1, $\boldsymbol{b}_p = \frac{1}{3}[\bar{1}12]$ $\boldsymbol{g} \cdot \boldsymbol{b}_p$	Dislocation 4, $\boldsymbol{b}_p = \frac{1}{3}[1\bar{1}\bar{2}]$ $\boldsymbol{g} \cdot \boldsymbol{b}_p$
$[\bar{1}11]$	invisible	visible	$+\frac{4}{3}$	$-\frac{4}{3}$
$[111]$	visible	invisible	$+\frac{2}{3}$	$-\frac{2}{3}$
$[11\bar{1}]$	invisible	visible	$-\frac{2}{3}$	$+\frac{2}{3}$

By referring to table 4.18 it can be seen that $\boldsymbol{b}_p = \pm\frac{1}{6}[\bar{1}12]$ does not account for the contrast because, if $\boldsymbol{g} \cdot \boldsymbol{b}_p$ is $-\frac{2}{3}$ for $(\bar{1}11)$ reflection, invisibility should be obtained for all three reflections or if $\boldsymbol{g} \cdot \boldsymbol{b}_p = +\frac{2}{3}$ for $(\bar{1}11)$ the dislocations should only be visible in this case. However, if $\boldsymbol{b}_p = \frac{1}{3}[\bar{1}12]$ dislocation 1 and $\boldsymbol{b}_p = \frac{1}{3}[1\bar{1}\bar{2}]$ dislocation 4, $\boldsymbol{g} \cdot \boldsymbol{b}_p$ has the values in table 4.18 and using rules 1 and 2 above we obtain the observed visible/invisible behaviour.

Thus we may describe the situation by the line diagram in figure 4.25(h) and using the rule for the summation of $\boldsymbol{b}_p$ at a node we see that the Burgers vectors are conserved. This situation is somewhat idealised since it is assumed that there is no dissociation in the matrix and that dislocations 2–4 lie in the particle–matrix interface.

A study of all $\boldsymbol{b}_p$ in figure 4.25(h) leads to the conclusion that the leading stacking fault is intrinsic while the larger trailing stacking fault is extrinsic and this was confirmed by Kear *et al.* (1969) using analysis of fringe images obtained in foils taken perpendicular to the tensile axis so that the critical plane was inclined to the foil surface.

As pointed out in section 3.4.5 it is very important to realise that such an analysis of $\boldsymbol{b}_p$ for partial dislocations is very difficult. It is very dangerous to rely solely on rules 1 and 2 above because Clarebrough (1971) has pointed out that in some cases partial dislocations may be invisible if $\boldsymbol{g} \cdot \boldsymbol{b}_p = +\frac{2}{3}$ even when $w \gtrsim 0.7$, i.e. rule 1 does not always apply. Furthermore, a significant amount of judgement may be necessary in assessing whether a dislocation is visible or invisible, for example dislocation 3, figure 4.25(b), is visible. Finally, the above rule 2 that $\boldsymbol{g} \cdot \boldsymbol{b}_p = +\frac{4}{3}$ is invisible does not appear to be general. Tunstall (1969) finds strong contrast for Frank loops in aluminium for $\boldsymbol{g} \cdot \boldsymbol{b}_p = \pm\frac{4}{3}$ with $w \gtrsim 0.7$. As a result it is vitally important that cross-checks such as $\boldsymbol{b}_p$ conservation at a node and/or the extrinsic/intrinsic nature of the stacking fault be made to confirm the consistency of the result if the above approach is taken. In general, it is strongly advised to use computer image matching techniques if $\boldsymbol{g} \cdot \boldsymbol{b}_p = \pm\frac{2}{3}$ cases cannot be avoided.

4.11.3 *Determination of Interparticle Spacings in Dispersion Strengthened Ni–Cr Alloys (after B. A. Wilcox)*

This section describes the determination of particle size distributions and interparticle spacings for a number of Ni–Cr alloys containing approximately 1 wt% of ThO_2.

Figure 4.26 shows transmission electron micrographs of typical areas of Ni–20 wt% Cr and 30 wt% Cr alloys. A high dislocation density is present and the ThO_2 particles have a wide range of sizes, although all are visible showing dark absorption contrast arising from the high atomic number of thorium. If electron micrographs are obtained under kinematical conditions, that is $\boldsymbol{s}$ large, the dislocation contrast is faint and the dark particles are clearly visible on a uniformly light

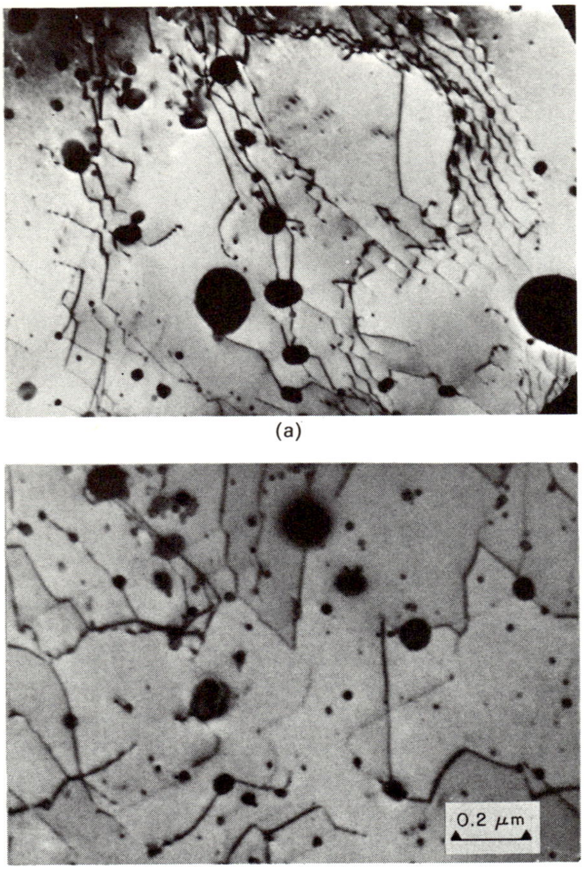

Figure 4.26 BF images of as-received alloys: (a) Ni–20 wt% Cr–1 wt% ThO_2; (b) Ni–30 wt% Cr–1 wt% ThO_2 [Courtesy of B. A. Wilcox]

background. A large number of photographs may be obtained, selecting areas randomly, and analysed using a Zeiss particle size analyser. A typical series of results taken from the work of Wilcox and Clauer (1969) are shown in figure 4.27. Here a minimum of 900–1000 particles have been measured per graph and it is clear that a wide distribution of particle size exists. This information may be converted into a mean planar separation d using the relation

$$d^2 = \frac{8}{3\sum_i f_i/r_{vi}^2} \tag{4.10}$$

where f_i is the volume fraction of the particles of a limited size range having an average radius r_{vi}. The results for the data in figure 4.27 are shown in table 4.19.

Alternatively, and less accurately, the mean planar spacing may be found using the relation (Ashby, 1964)

$$d^2 = \frac{4r_s^2}{f} \tag{4.11}$$

where r_s is the mean particle radius and f is the volume fraction. These results are also shown in table 4.19 and are clearly significantly in error.

4.12 Steels

Steels are the single, most extensively used structural material and have been studied in detail using transmission electron microscopy. Many alloy steels have precipitation hardened microstructures and it is from this class of materials that most of the examples in this section are taken.

4.12.1 *Identification of $Cr_{23}C_6$ Precipitates at Grain Boundaries in Stainless Steel (after P. H. Pumphrey)*

Austenitic stainless steels are frequently strengthened with carbide precipitates and one of the most common is $Cr_{23}C_6$. This section describes a method for the identification of $Cr_{23}C_6$ in a 25 wt% Ni–20 wt% Cr–0.5 wt% C austenitic stainless steel. Figure 4.28 shows grain boundary precipitate after the material had been recrystallised for 15 min at 1050 °C then aged at 700 °C for 30 min. The precipitate particles P are growing

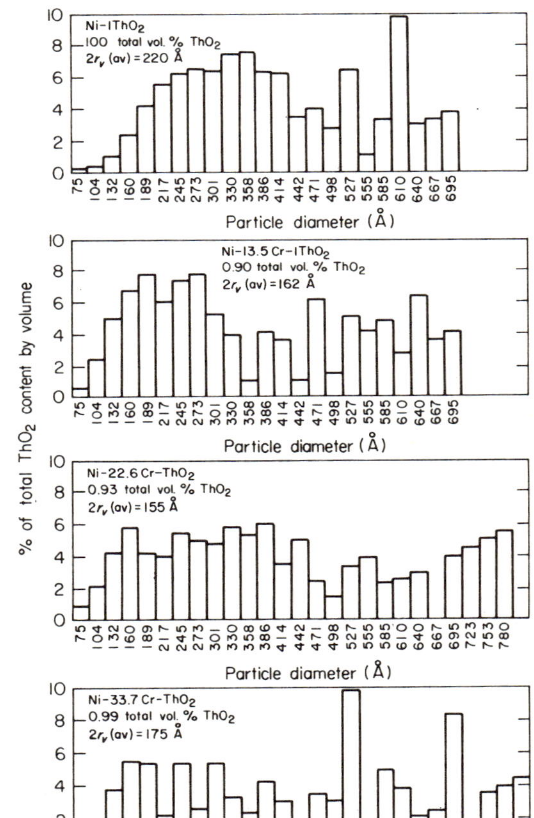

Figure 4.27 Particle size distributions from thoriated nickel-base alloys [From Wilcox and Clauer (1969), courtesy of *Metal Science Journal*]

Table 4.19 Composition and ThO_2 particle sizes and spacings

Alloy	Wt% Cr	Wt% ThO_2	Vol% ThO_2	Nominal range of particle diameter (Å)	Average particle diameter $2r_v$ (Å)	Mean planar centre-to-centre particle spacing d (Å)	
						Equation (4.10)	Equation (4.11)
Ni–1 ThO_2	0	1.12	1.0	75–695	220	2505	2200
Ni–13.5 Cr–1 ThO_2	13.5	1.05	0.90	75–695	162	2070	1707
Ni–22.6 Cr–1 ThO_2	11.6	1.10	0.93	75–780	155	2165	1607
Ni–33.7 Cr–1 ThO_2	33.7	1.20	0.99	75–810	175	2290	1758

by the migration of the incoherent interface with grain 2 (Singhall and Martin, 1968) and are semi-coherent with grain 1.

The approach taken to identify the particles is that outlined in section 2.13.1. SADP were taken with simple $\boldsymbol{B}$ in the matrix of grain 1 but including several precipitates within the SAD aperture. The results are shown in figure 4.29(a)–(c) in which the strong spots are from the matrix while the weak ones are from the precipitates. The arrowed spots arise from the small part of grain 2 included within the SAD aperture. The following conclusions may be drawn.

(1) All carbide particles have parallel orientation because simple arrays of carbide spots are obtained.

(2) The particles have almost an exact orientation relationship with the matrix because complete zones of carbide spots are obtained.

Using the method of indexing outlined in section 2.7.2 combined with the matrix lattice parameter $a = 3.59$ Å and the camera constant, the matrix patterns were indexed as shown in figure 4.30(a)–(c), that is $\boldsymbol{B}$ in the matrix is [001], [013], [011]. This result was consistent with the angles of tilt measured from the tilt controls of the microscope, that is [001] → [013] 18.5°, [013] → [011] 26.5°, about OA, $\boldsymbol{g}$[200]. The crystal structure of the precipitate may be obtained by determining the ratios of the interplanar spacings associated with the spots in the SADP, that is the ratios of the spot spacings, see section 2.7.2.1. This information is summarised in table 4.20. Also shown are inter-spot distances squared and subsequently doubled which are proportional to 3, 4, 8, 11, that is the material is f.c.c., see table 2.2. Thus the spot type may be indexed as shown in table 4.20 and the lattice parameter, obtained using the equations in table 2.2, is 10.62 Å which corresponds to $Cr_{23}C_6$

Figure 4.28 BF image of $Cr_{23}C_6$ precipitates in an austenitic stainless steel direct quenched from 1050 °C and aged at 700 °C for 30 min [Courtesy of P. H. Pumphrey]

Figure 4.29 SADP taken from the particles in figure 4.28 with the specimen tilted to (a) $\boldsymbol{B}$[100]; (b) $\boldsymbol{B}$[013]; (c) $\boldsymbol{B}$[011] matrix

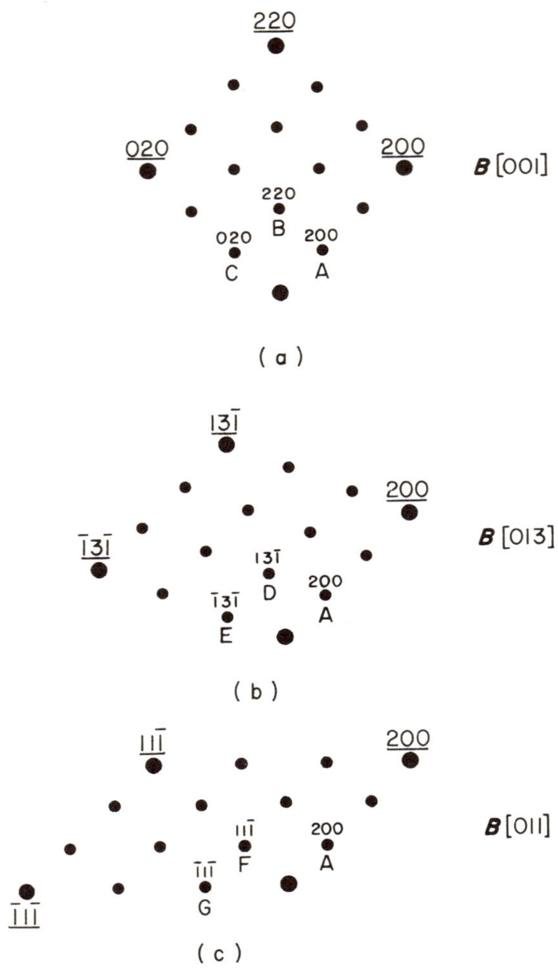

Figure 4.30 (a)–(c) Indexed patterns for figure 4.29(a)–(c)

(Andrews *et al.*, 1971). The $Cr_{23}C_6$ spots may be indexed as shown in figures 4.30(a)–(c). Clearly, the orientation relationship between precipitate and matrix is cube–cube. Note that the operative matrix reflection in figure 4.28 is {111} and that the Moiré pattern corresponds to a parallel misfit pattern because the fringes are perpendicular to $\boldsymbol{g}$, see section 3.21.1. Using equation (3.25) which relates the interplanar spacings to the fringe pattern spacing, it may be shown that the interplanar misfit is 0.986% which is consistent with the measured lattice parameter difference, that is $3\times a(\text{matrix}) = 10.77$ Å, compare 10.62 Å for $Cr_{23}C_7$.

Table 4.20 Interspot distances R, R^2 and $2R^2$ measured in centimetres on a photographic enlargement

Distances	R	R^2	$2R^2$	hkl
OA	1.46	2.13	4.26	{200}
OB	2.02	4.08	8.16	{220}
OC	1.42	2.02	4.04	{200}
OD	2.34	5.47	10.94	{113}
OE	2.34	5.47	10.84	{113}
OF	1.23	1.51	3.02	{111}
OG	1.23	1.51	3.02	{111}

4.12.2 *Identification of Precipitates in Nitrogen-containing Stainless Steel* (*after D. W. Borland*)

An alloy containing 18 wt % Cr, 12.1 wt % Ni, 1.23 wt % Nb and 0.09 wt % N was solution treated at 1300 °C for 1 hour, water quenched and then aged at 700 °C. Copious stacking fault precipitation occurred. This section describes the identification of the precipitating phase from extraction replicas. Figure 4.31(a) shows a typical carbon extraction replica, the sheets of precipitate corresponding to the stacking fault precipitate described by Silcock and Tunstall (1964). Figure 4.31(b) shows a diffraction pattern obtained with a large diffraction aperture which included a very large number of precipitate particles from a number of grains. As a result the pattern consists of the familiar rings. Two phases are present and the arrowed rings from one only are of interest for present purposes. Table 4.21 contains ring diameter measurements which have been squared and trebled. These latter values are proportional to 3, 4, 8, 11, etc., and the material is clearly f.c.c., see table 2.2. Thus the spot types may be indexed as shown and the lattice parameter may be calculated using the equations in table 2.2 to give the values shown in table 4.21. The average lattice parameter is 4.371 Å which agrees within the experimental error with that for one form of NbN quoted by Goldschmidt (1967) as $a = 4.381$ Å.

Table 4.21 Electron diffraction data from the original plates of figure 4.31(b)

Ring diameter (cm)	D^2 (cm^2)	$3D^2$ (cm^2)	$\{hkl\}$	a (Å)
1.01	1.01	3.03	{111}	4.365
1.16	1.35	4.05	{200}	4.490
1.66	2.75	8.25	{220}	4.335
1.94	3.75	11.25	{311}	4.354
2.02	4.1	12.3	{222}	4.363
2.32	5.4	16.2	{400}	4.390
2.52	6.4	19.2	{331}	4.399

Camera constant $2\lambda L = 25.45$ cm Å.

4.12.3 *Indexing a Complex Diffraction Pattern from an Alloy Steel—Confirmation of the Presence of VC and its Orientation Relationship to the Matrix* (*after G. L. Dunlop*)

This example of the method involves indexing a complex spot pattern from a transformed, that is ferritic, steel containing vanadium carbide precipitates and having an oxide film in the surface

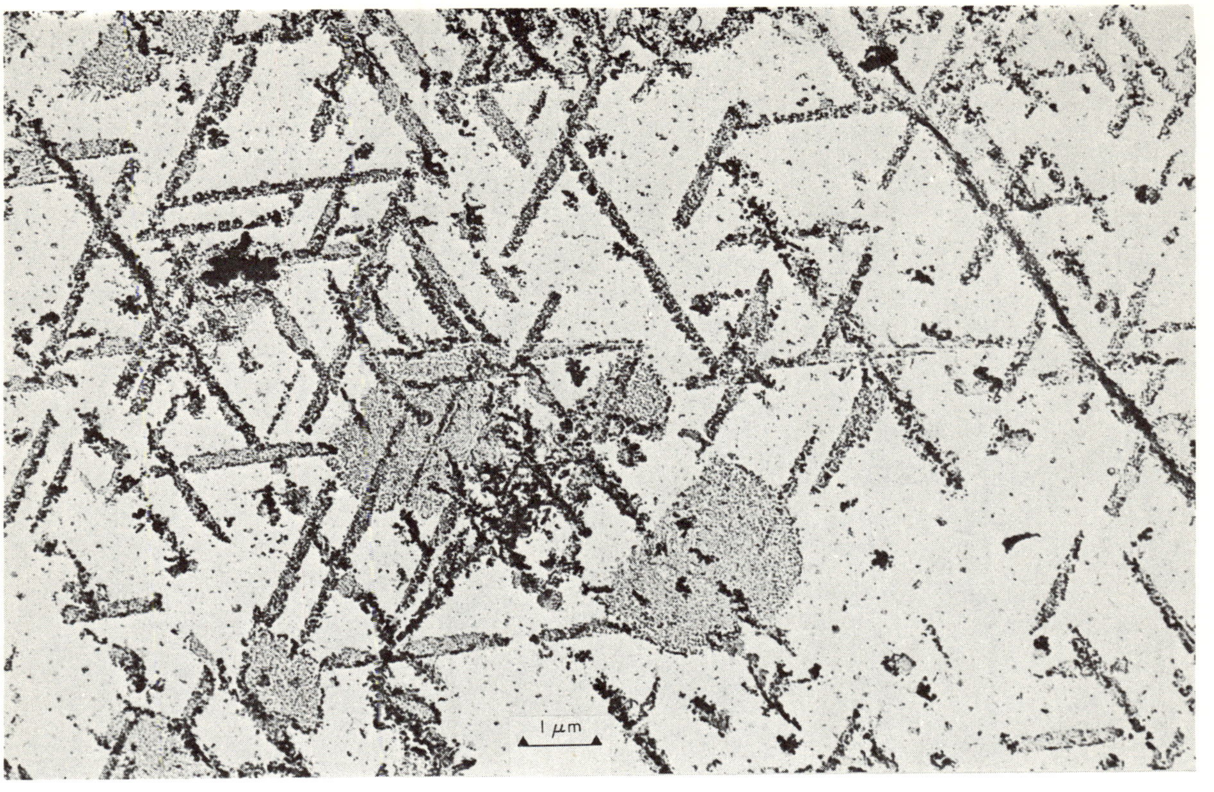

(a)

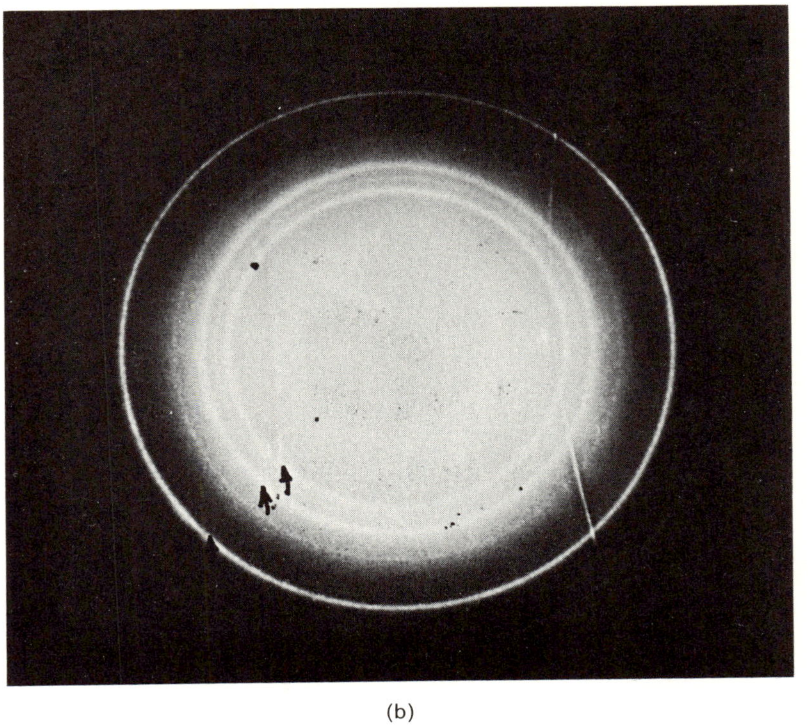

(b)

Figure 4.31 An extraction replica from an austenitic stainless steel (18 wt % Cr–12.1 wt % Ni–1.23 wt % Nb–0.09 wt % N) aged for 200 hours at 700 °C showing stacking fault precipitate. (b) Diffraction pattern from (a) [Courtesy of D. W. Borland]

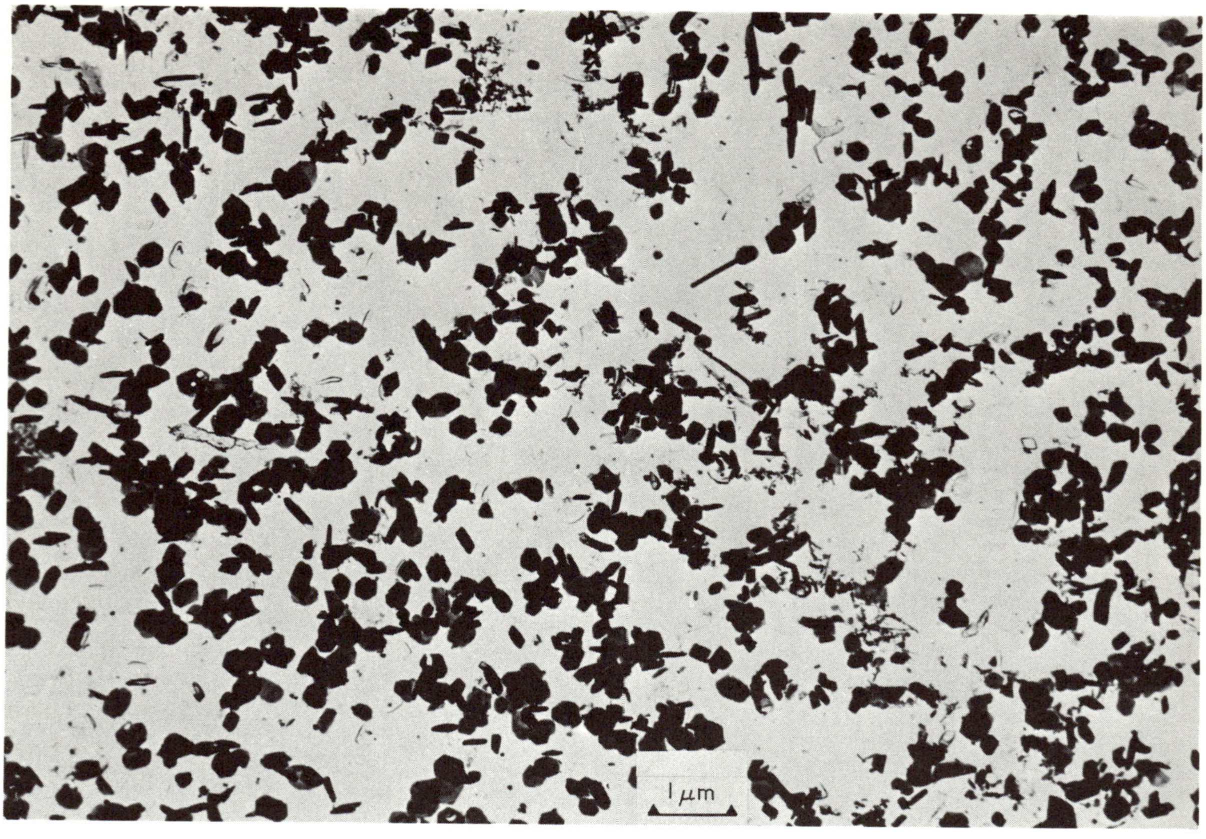

(a)

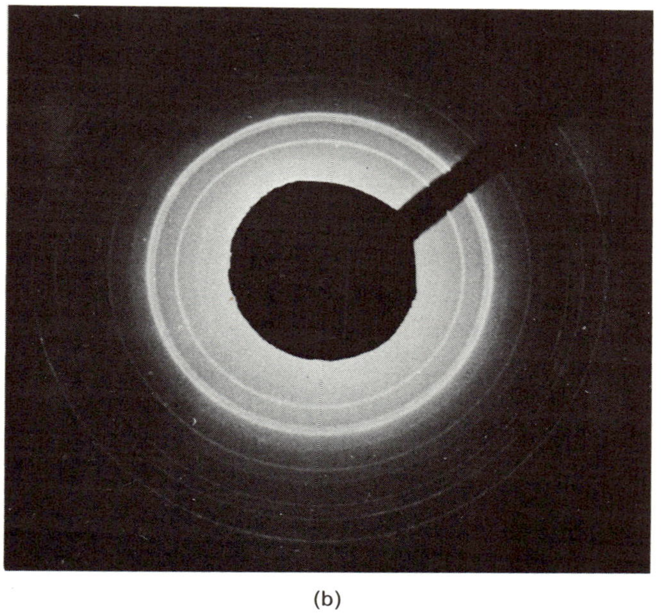

(b)

Figure 4.32 (a) An extraction replica from an austenitic stainless steel (17.9 wt % Cr–12.1 wt % Ni–1.25 wt % Nb–0.05 wt % N) aged for 800 hours at 700 °C. (b) Diffraction pattern from (a) [Courtesy of D. W. Borland]

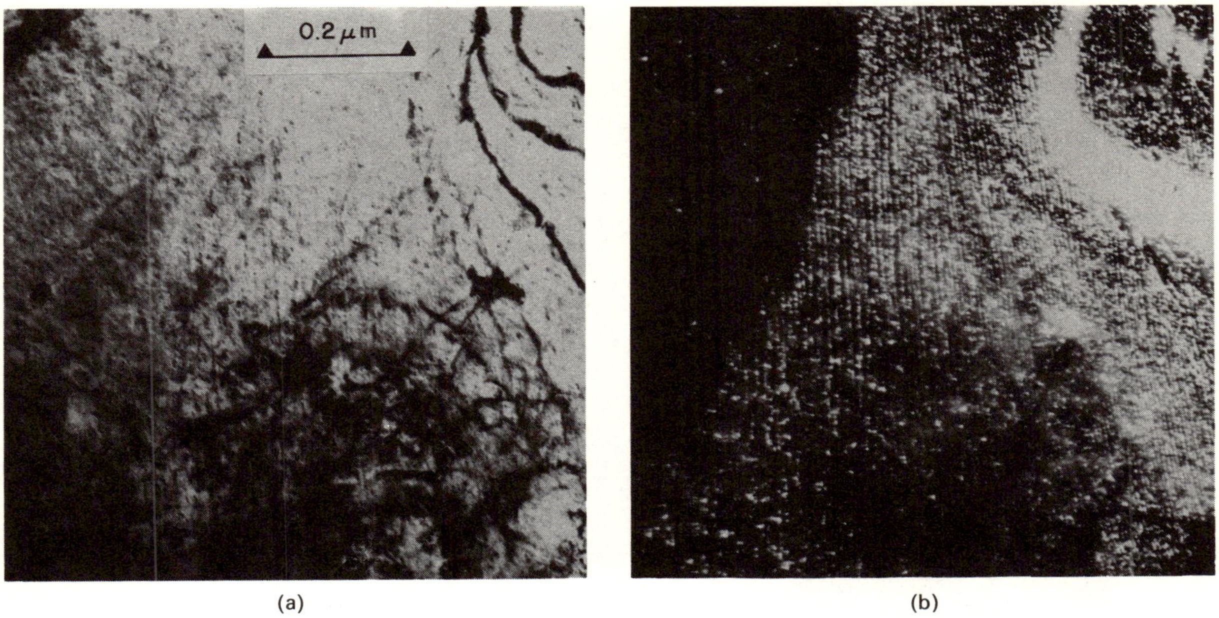

Figure 4.33 (a) BF and (b) CDF using a vanadium carbide reflection from a steel containing 0.32 Cr, 0.51 Mo, 0.24 V, 0.59 Mn, 0.11 Ni, 0.25 Si, 0.12 Cu, 0.029 S, 0.015 P, 0.10 C in wt%, isothermally transformed at 700 °C for 15 min [Courtesy of G. L. Dunlop]

of the thin foil. The composition of the steel is 0.32 wt % Cr, 0.51 wt % Mo, 0.24 wt % V, 0.59 wt % Mn, 0.11 wt % Ni, 0.25 wt % Si, 0.12 wt % Cu, 0.029 wt % S, 0.015 wt % P and 0.10 wt % C and it was solution treated at 1150 °C and then isothermally transformed at 700 °C for 15 min. This alloy contains the interphase carbide precipitate in an α-iron matrix with the row morphology described by Davenport *et al.* (1968) and by Davenport and Honeycombe (1971), see figure 4.33(a),(b).

Two SADPs were obtained, see figure 4.34(a),(b), from different areas with different ***B***. Using the method of ratios and the rules of indexing in section 2.7.2.1, the matrix patterns (strong spots marked X) may be indexed as $\boldsymbol{B}[110]_\alpha$ and $[100]_\alpha$ as shown in figure 4.34(c),(d). The remaining features of the SADP unaccounted for are a number of spots and a series of faint rings. These spots and rings will be considered separately.

4.12.3.1 *Indexing the extra spots,* $\boldsymbol{B}[110]_\alpha$

The distances of the extra spots from the centre spot (figure 4.34(a)) may be converted into d spacings using the camera constant λL (equation (1.11)) obtained from measurements performed on the matrix α-iron pattern. In the case of arced reflections the radius of the arc is measured. Using the lattice parameters of $VC_{0.95}$ as equal to 4.160 Å and Fe_3O_4 as 8.392 Å (f.c.c.), the reflections may be indexed as shown in table 4.22.

By arbitrarily indexing spot A as (020) and choosing spot B as (022), the correct interplanar angles are obtained and $\boldsymbol{B}[100]_{VC}$ with the remaining spots indexed as shown in figure 4.35(a). Spot E may be arbitrarily indexed as $(02\bar{2})$, spot F as $(20\bar{2})$ Fe_3O_4 and $\boldsymbol{B}[111]$ Fe_3O_4 is obtained with full indexing as shown in figure 4.35(b). Spot G may be arbitrarily chosen to be $(\bar{1}3\bar{1})$ and spot H is $(02\bar{2})$ Fe_3O_4 which gives $\boldsymbol{B}[211]_{Fe_3O_4}$ and the

Table 4.22 Measured interplanar spacings for the spots in figure 4.34(a) and calculated interplanar spacings for $VC_{0.95}$ and Fe_3O_4

Spot designation figure 4.34(a)	Interplanar spacing (Å) (calculated from figure 4.34(a)), original plates	Interplanar spacing $VC_{0.95}$		Fe_3O_4	
		hkl	*d* (Å)	*hkl*	*d* (Å)
A	2.10	{002}	2.080		
B	1.48	{022}	1.471		
C	2.10	{002}	2.080		
D	2.98			{022}	2.967
E	2.98			{022}	2.967
G	2.55			{131}	2.530
F	2.98			{022}	2.967

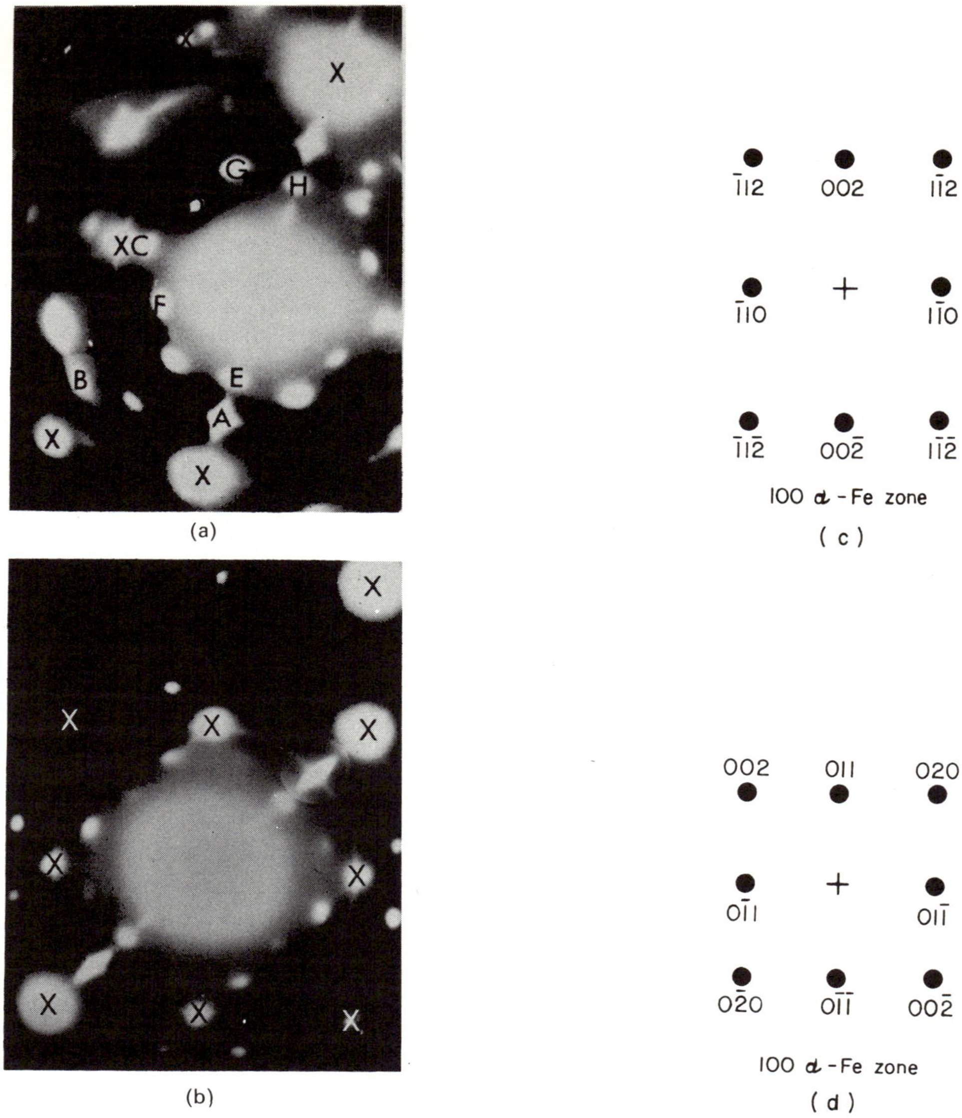

Figure 4.34 SADP (a), (b) taken from the same material shown in figure 4.33. (c) and (d) are indexed matrix zones corresponding to (a), (b) [Courtesy of G. L. Dunlop]

pattern shown schematically in figure 4.35(c). The ($\bar{1}11$) and ($1\bar{1}\bar{1}$) reflections are obscured by the inelastic scattering near the transmitted beam. Thus all the spots are accounted for and the interpretation is as follows.

(1) $VC_{0.95}$ is present with the orientation relationship

$$[100]_{VC}//[110]_{\alpha\text{-iron}}$$

$$(001)_{VC}//(001)_{\alpha\text{-iron}}$$

obtained directly from the indexed diffraction patterns, figure 4.35(d). Reference to the stereogram in appendix 9 shows that this is the Baker–Nutting (1959) orientation relationship previously reported by Tekin and Kelly (1965) and by Davenport (1968) for VC in α-iron.

(2) Fe_3O_4 is present on both surfaces of the foil arising from oxidation of the thin foil. The oxide on the top and bottom surfaces is responsible for the two patterns indexed schematically in figure 4.35(b),(c) and has the orientation relationship

figure 4.35(b) $[111]_{Fe_3O_4}//[110]_\alpha$

$$(0\bar{1}1)_{Fe_3O_4}//(001)_\alpha$$

figure 4.35(c) $[211]_{Fe_3O_4}//[110]_\alpha$

$$(0\bar{1}1)_{Fe_3O_4}//(001)_\alpha$$

obtained directly from the indexed diffraction pattern.

Reference to the stereogram in appendix 9 shows that these correspond to the variants of the Nishiyama–Wasserman orientation relationship which has previously been reported for Fe_3O_4–α-iron (Keown and Dyson, 1966). This can easily

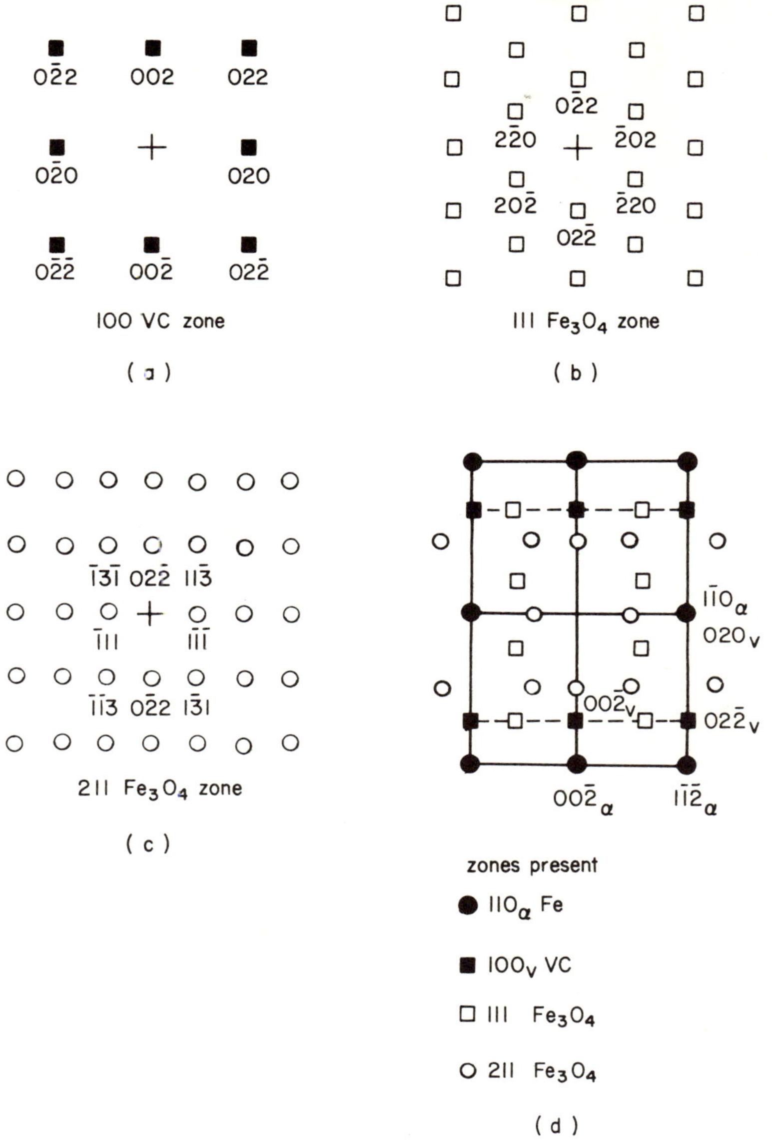

Figure 4.35 (a) $VC_{0.95}$ and (b), (c) Fe_3O_4 indexed patterns. (d) Superimposed patterns from figure 4.34(a) [Courtesy of G. L. Dunlop]

be seen by checking which poles in Fe_3O_4 can be parallel for all types of $\boldsymbol{B}\langle 110\rangle_{\alpha\text{-iron}}$ and by determining which planes of α and Fe_3O_4 are parallel on all $\langle 110\rangle_{\alpha\text{-iron}}$ (great circle).

The above indexing of the SADP may be demonstrated to be correct using CDF images. For example, figure 4.33(b) shows CDF of the region from which the SADP in figure 4.34(a) was taken. The $(0\bar{2}\bar{2})$ carbide spot B, figure 4.34(a), was used for the CDF and the rows of carbide particles characteristic of interphase precipitation (Davenport and Honeycombe, 1971) are clearly visible.

It is left as an exercise to show that the indexed patterns in figure 4.36 describe the spots in figure 4.34(b), using the Baker–Nutting and different variants of the Nishiyama–Wasserman orientation relationships.

4.12.3.2 *Indexing the faint ring*

In both SADPs the rings run exactly through VC spots (figure 4.35(a),(b)) and are therefore considered to arise from randomly oriented VC particles that have remained attached to the foil surface after the matrix surrounding them has been polished away during preparation of the specimen. They can therefore be given the same indices as the spots.

4.12.3.3 *Detection of VC in* α-iron *when* Fe_3O_4 *is present*

It is a common problem in α-iron alloys to have carbide particles present in foils that are slightly oxidised. Thus both Fe_3O_4 and VC reflections are present in the SADP and these may be used for

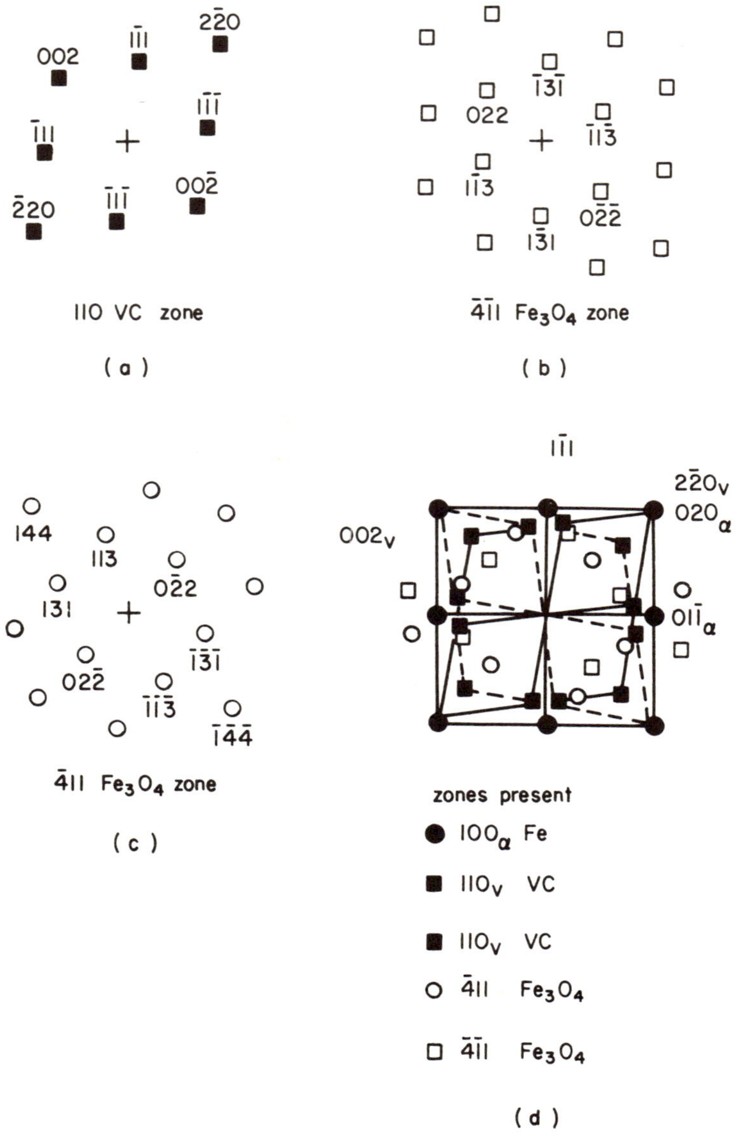

Figure 4.36 (a) $VC_{0.95}$ and (b), (c) Fe_3O_4 indexed patterns. (d) Superimposed patterns from figure 4.34(b) [Courtesy of G. L. Dunlop]

CDF imaging to make the VC particles or oxide particles visible independently. However, sometimes it may be difficult to identify uniquely spots in the diffraction pattern because the VC and Fe_3O_4 may take up specific variants of their respective orientation relationships such that their diffraction spots superimpose or are reinforced by double diffraction from VC to Fe_3O_4 reflections and vice versa. Davenport (1968) has shown that the distinction between $VC_{0.75}$ and Fe_3O_4 can be extremely difficult during the early stages of precipitation for particular combinations of VC and Fe_3O_4 variants. For example, if the variants combine to give $\boldsymbol{B}\langle 111\rangle_{VC}$ and $\langle 111\rangle_{Fe_3O_4}$ we have a situation which is summarised in the diagrams in figure 4.37. The $\{220\}_{VC_{0.75}}$ reflections are streaked parallel to $\langle 100\rangle$, see figure 4.37(a) for precipitate plates on $\{100\}$. If $\boldsymbol{B}$ is exactly $[011]_{\alpha\text{-iron}}$ the carbide spot positions almost coincide with those of the matrix and Fe_3O_4, see figure 4.37(b). However, by careful tilting about $[011]_{\alpha\text{-iron}}$ the (220) $VC_{0.75}$ streak and the matrix spot $(200)_{\alpha\text{-iron}}$ can be separated, see figure 4.37(c),(d), and centred DF performed. The nearby oxide streak disappears on tilting (Davenport, 1968).

If the carbide is a different orientation variant, separation of oxide and carbide using CDF may be easier, see figure 4.34(a).

In practice, each case must be treated on its merits. The aim should be to attempt to obtain SADP in which spots from Fe_3O_4 on both surfaces are present and may be uniquely identified relative to the $VC_{0.75}$ spots. After checking for possible contributions to $VC_{0.75}$ spots by double diffraction from α-iron and Fe_3O_4 spots CDF may be per-

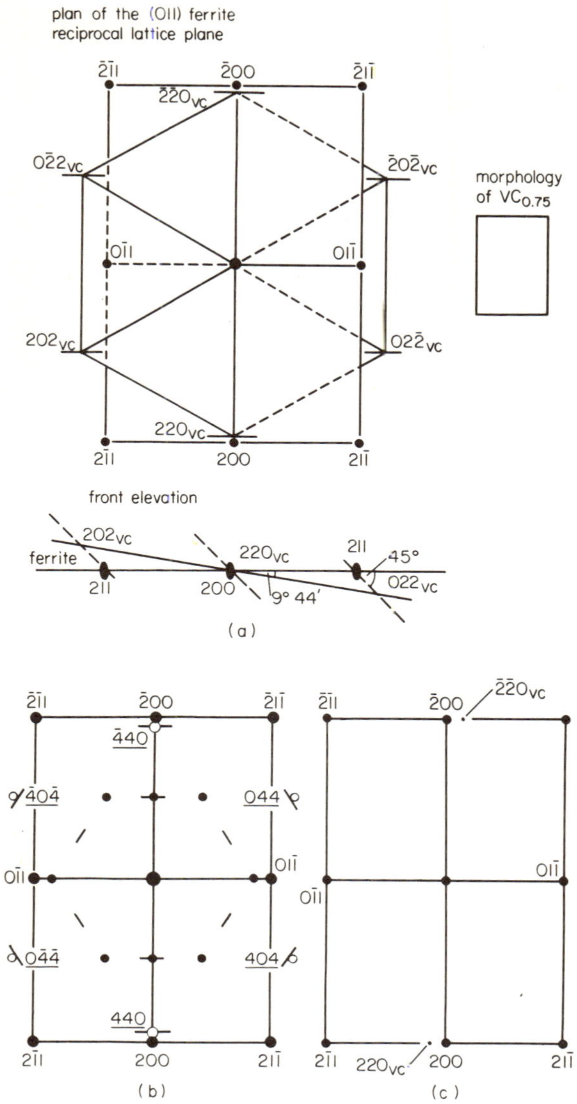

Figure 4.37 (a) Diagram showing the relationships between the reciprocal lattice points for $\boldsymbol{B}[011]_{\alpha\text{-iron}}$ and $[\bar{1}11]_{VC_{0.75}}$. (b) Schematic diagram showing position of $VC_{0.75}$ spots if $\boldsymbol{B}[011]_{\alpha\text{-iron}}$. (c) As for (b) but $\boldsymbol{B} \sim [011]_{\alpha\text{-iron}}$

formed using the carbide spots. A cross-check may be obtained if the morphology or distribution of the precipitate is different from that of the oxide, for example figure 4.33(b).

4.12.4 *Determination of the Orientation Relationship Between Ferrite and Cementite in a Pearlite Colony (after R. J. Dippenaar)*

The orientation relationship between ferrite and cementite in pearlite has been studied by a number of workers and is generally either of the type

$$(100)_C//(0\bar{1}1)_\alpha,\ (010)_C//(1\bar{1}\bar{1})_\alpha,\ (001)_C//(211)_\alpha$$

generally known as the Bagaryatski (1950) relationship, or

$$(100)_C\ 2.6^\circ \text{ from } (13\bar{1})_\alpha,\ (010)_C\ 2.6^\circ \text{ from } (113)_\alpha$$

$$(001)_C//(5\bar{2}\bar{1})_\alpha$$

generally known as the Pitsch (1962) relationship. An example is shown here of the determination of the Pitsch relationship in a pearlite colony in a 12 wt% Mn and 0.8 wt% C steel.

As pointed out in section 2.7.2.1, SADPs obtained in the electron microscope are always centrosymmetric, apart from variation in the intensities of the spots. Consequently, the electron diffraction pattern cannot be uniquely indexed because there are two possibilities related by 180° rotation about $\boldsymbol{B}$ (180° ambiguity). In section 2.8.1 a number of methods were described for removing the 180° ambiguity by using $\boldsymbol{B}$ with reflections from more than one zone present in the SADP (Ryder and Pitsch, 1968). However, in pearlite it is often difficult to fulfil this requirement because tilt conditions must be found such that both crystals give interpretable diffraction patterns and at the same time reflections must be present from two zones in both phases. The situation is further complicated by the ambiguity in interpreting cementite diffraction patterns (see appendix 14). Nevertheless, it is still possible to index the diffraction patterns uniquely using the following approach.

(1) Select a tilting condition such that the cementite gives a diffraction spot pattern of uniform intensity and enough ferrite reflections are present to form an interpretable diffraction pattern.

(2) Record the SADP.

(3) Tilt the foil through $\sim +20^\circ$ until a second symmetrical cementite zone is obtained which satisfies the conditions in (1).

(4) Record the diffraction pattern.

(5) If possible tilt through $\sim -20^\circ$ from the original position to obtain a third symmetrical SADP.

(6) Record the diffraction pattern.

(7) Index the SADP from both phases in order to obtain a consistent orientation relationship.

An example of this approach is now shown using figure 4.38(a),(d) which were obtained from a pearlite colony with a 20° specimen tilt applied between recording the two SADPs. Figures 4.38(b),(c),(e),(f) show alternative methods of indexing these patterns taking into account the 180° ambiguity. We may now analyse these patterns using a stereogram in the following way.

(1) Draw the diffracting zone of ferrite on a standard (001) cubic stereogram.

(2) Draw the diffracting zone of cementite on a standard (001) cementite stereogram.

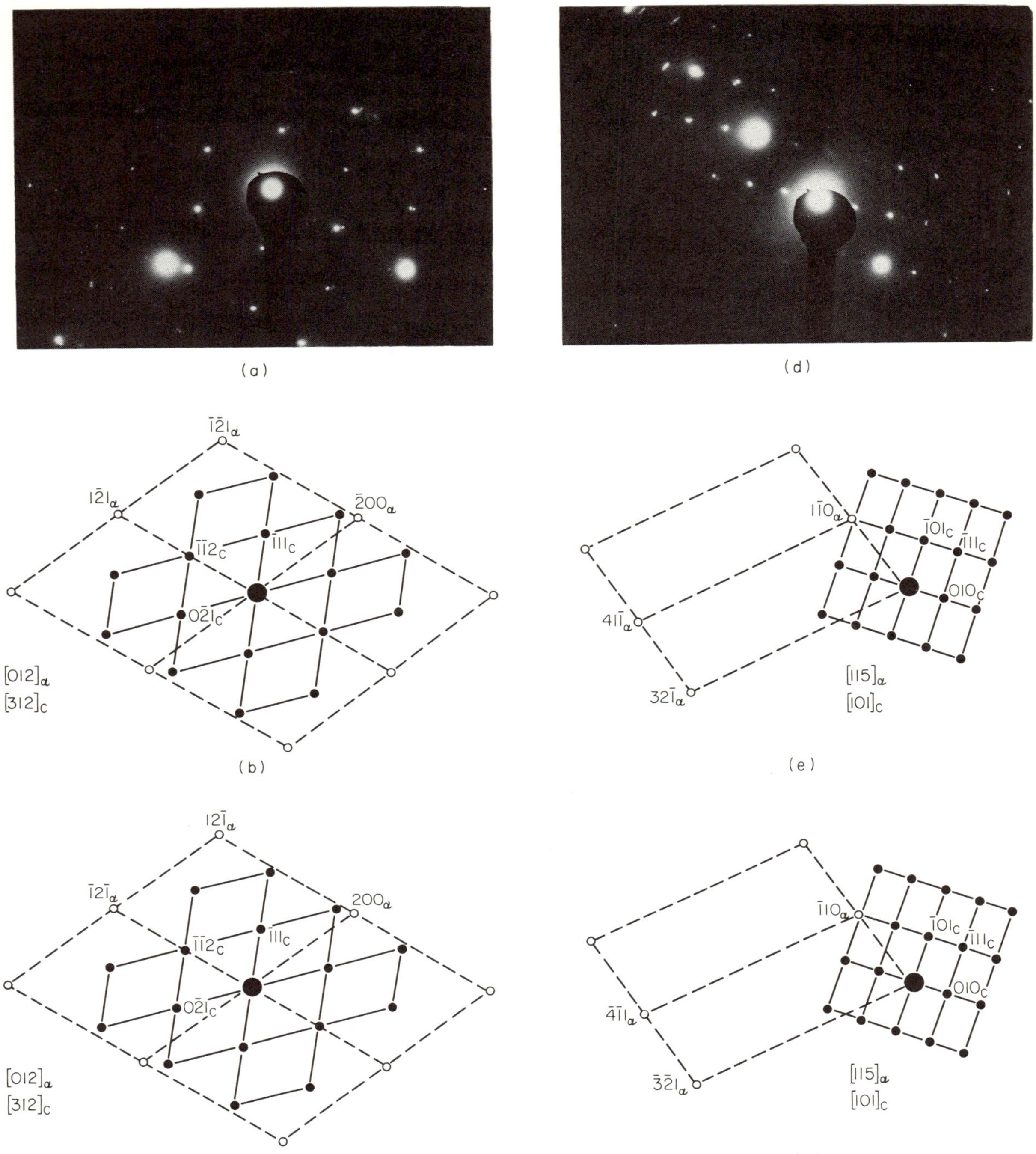

Figure 4.38 Fe–12 wt % Mn–0.8 wt % C: (a) diffraction pattern from a pearlite area, (b) first and (c) second possibility of indexing the two reflecting zones; (d) diffraction pattern from pearlite; (e) first and (f) second methods of indexing

(3) Superimpose these two stereograms in such a way that, when the poles of the respective reflecting zones are projected to the centre of the superimposed stereogram, the angular relationship between the ferrite and cementite zones is maintained in each SADP, that is figure 4.38(a) $(1\bar{2}1)_\alpha//(\bar{1}\bar{1}2)_C$.

In figure 4.39(a) the $(1\bar{2}1)_\alpha$ and $(\bar{1}\bar{2}1)_\alpha$ reflections are shown on the great circle of the $[012]_\alpha$ zone is then projected to the centre of the stereogram and the $(1\bar{2}1)_\alpha$ and $(\bar{1}\bar{2}1)_\alpha$ positions project to the positions $(1\bar{2}1)_{\alpha p}$ and $(\bar{1}\bar{2}1)_{\alpha p}$ on the principal circle. Similarly, when the [312] cementite pole is projected to the centre of the cementite stereogram, the $(\bar{1}\bar{1}2)_C$ and $(\bar{2}03)_C$ reflections on the great circle of the cementite zone project to the positions $(\bar{1}\bar{1}2)_{Cp}$ and $(\bar{2}03)_{Cp}$ on the principal circle. The standard ferrite and cementite stereograms were rotated, as shown, to maintain the correct angular relationship between ferrite and cementite reflections, $(\bar{1}\bar{1}2)_C//(1\bar{2}1)_\alpha$. When the cementite pole is projected to the centre of the stereogram, the $\{100\}$ cementite poles are also projected through the same angle to the positions A, B and C on the stereogram. We now have essentially a (012) ferrite stereogram and a (312) cementite stereogram superimposed in the correct angular relationship and one can proceed to measure the angles between the principal poles of cementite and the principal poles of ferrite (if the principal ferrite poles were also projected on to the $(012)_\alpha$ stereogram). Then we have the following.

(4) The cementite stereogram is now rotated through 180° (that is $(\bar{1}2\bar{1})_\alpha//(\bar{1}\bar{1}2)_C$ instead of $(1\bar{2}1)_\alpha//(\bar{1}\bar{1}2)_C$) and the new positions of the principal cementite poles are marked D, E and F in this case. We now have the two possible positions of each principal cementite pole with respect to ferrite, but this is still on a $(012)_\alpha$ stereogram.

(5) In order to present the orientation relationship on a standard ferrite stereogram the (012) ferrite pole is projected back to its original position on the (001) cubic stereogram. The principal cementite poles are then rotated through the same angle to positions A′, B′, C′ for the first possibility of the orientation relationship to D′, E′ and F′ for the second possibility.

(6) The same procedure is followed for the diffraction pattern in figure 4.38(e),(f) and two possible orientation relationships are also found in this case, see figure 4.39(b).

(7) The two standard cubic stereograms, figure 4.39(a),(b), are then superimposed and it is immediately evident which two of the four possible relationships coincide, that is A′B′C′. This is then the real orientation relationship which may be described by reference to the standard (001) cubic projection as shown in figure 4.39(c) as

$$(100)_C \text{ close to } [131]_\alpha$$
$$(010)_C \text{ close to } [\bar{3}11]_\alpha$$
$$(001)_C \text{ close to } [1\bar{2}5]_\alpha$$

which is a variant of the Pitsch relationship.

It can clearly be seen that the errors arise

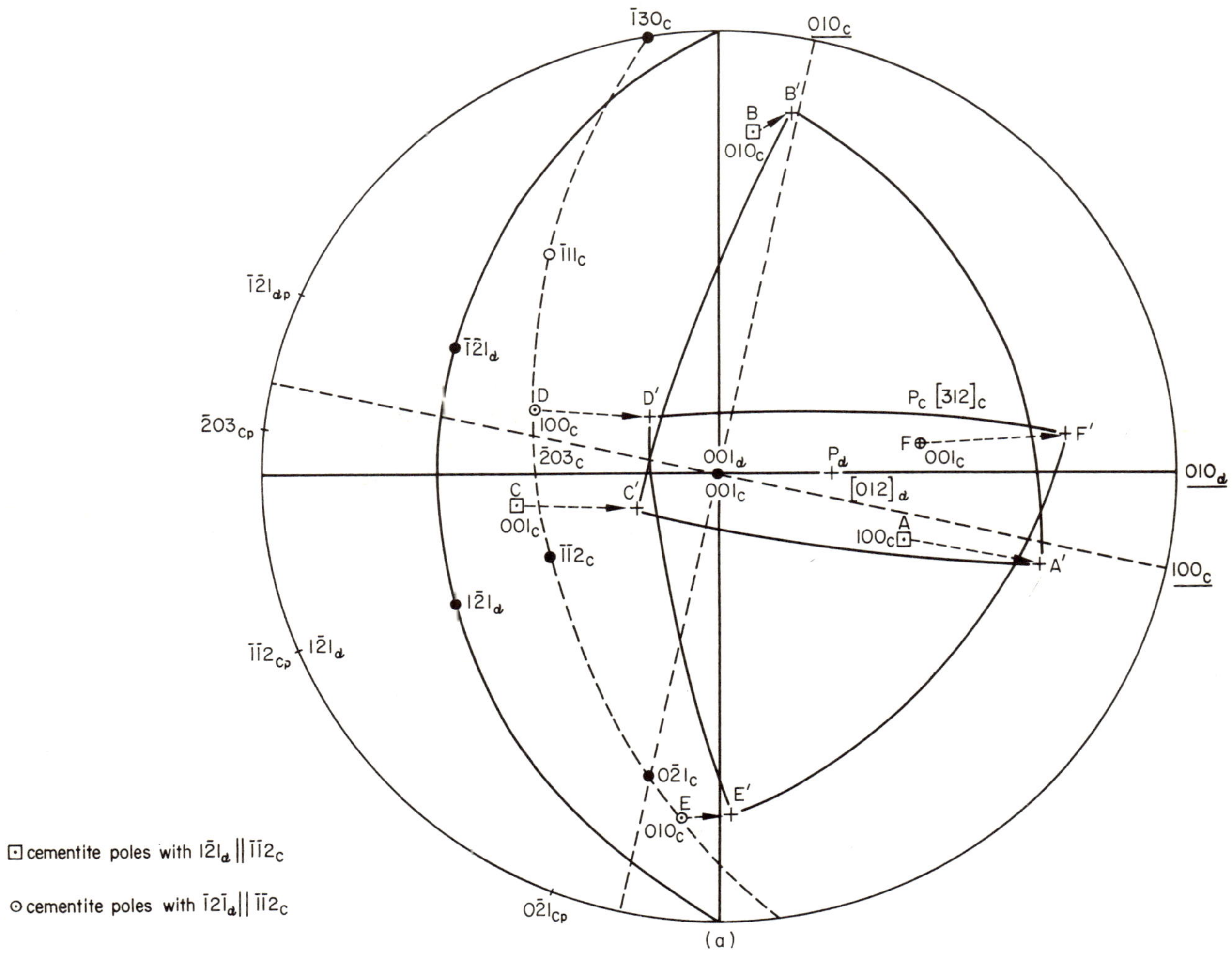

Figure 4.39 (a), (b) Stereograms describing the analysis of the SADP in figure 4.38 together with the final analysis (c)

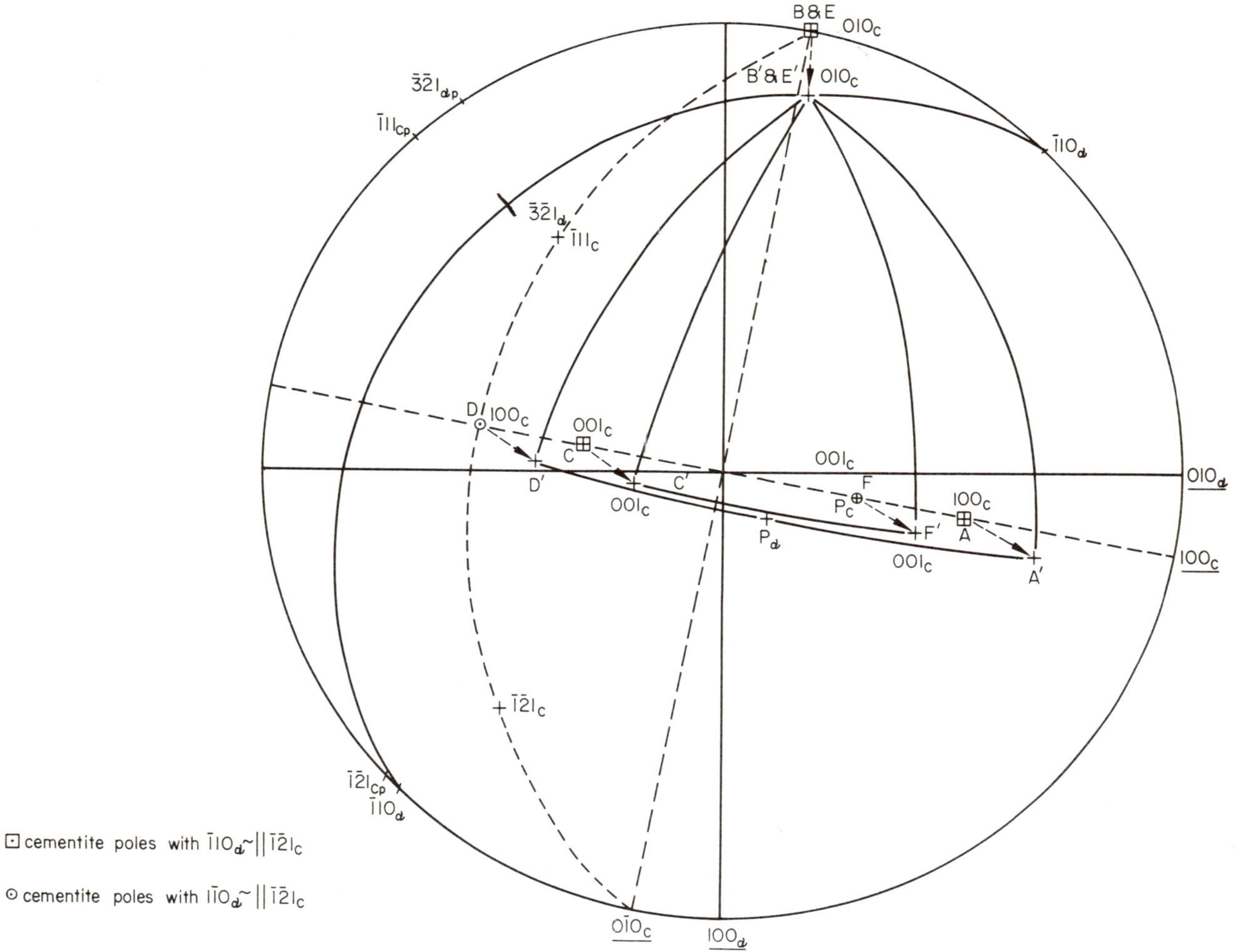

(b)

Figure 4.39 (*continued*)

primarily from inaccuracy in the orientation relationship determined from the diffraction pattern in figure 4.38(a), which has an asymmetric distribution of intensities of the spots, that is $\boldsymbol{B}$ is not exact.

This method is quite general and may be applied to all cases where two or more phases are present, for example Dippenaar and Honeycombe (1973). The accuracy of the orientation relationship determined using the stereographic projection is only $\pm 4°$. However, this may be improved if the analysis is computerised.

4.13 Titanium Alloys

The microstructures of many commercial titanium and zirconium alloys are similar, consisting of solution strengthened α phase (c.p.h.), α–β phase mixtures (β is b.c.c.) or transformed β. In this section examples are confined to titanium alloys but similar problems are also found in zirconium alloys. Examples are described of the identification of (1) deformation twins, (2) ω phase which is a common transformation product and (3) hydrogen contamination arising from electropolishing which is a frequent problem with these materials.

4.13.1 *Identification of an Unknown Twin in Titanium (after P. G. Partridge)*

The approach in this particular case used to identify this twin is based on the information in section 2.12.2.2. A SADP from the twin and matrix at the same angle of specimen tilt is shown in figure 4.40(a),(b). The matrix pattern was indexed and projected as a great circle in figure 4.40(c). The great circle corresponding to the trace of the pole of K_1 is also shown. This great circle indicates two reciprocal twins, $(10\bar{1}2)$ and $(\bar{1}012)$, which would give rise to the observed twin boundary trace. By superimposing the $(10\bar{1}2)$ and $(\bar{1}012)$ twin pole projections on (0001), that is by rotating the (0001) projection about $(\bar{1}2\bar{1}0)$ by $\pm 95°$ (table 2.11), it was found that the $(10\bar{1}2)$ twin poles agreed

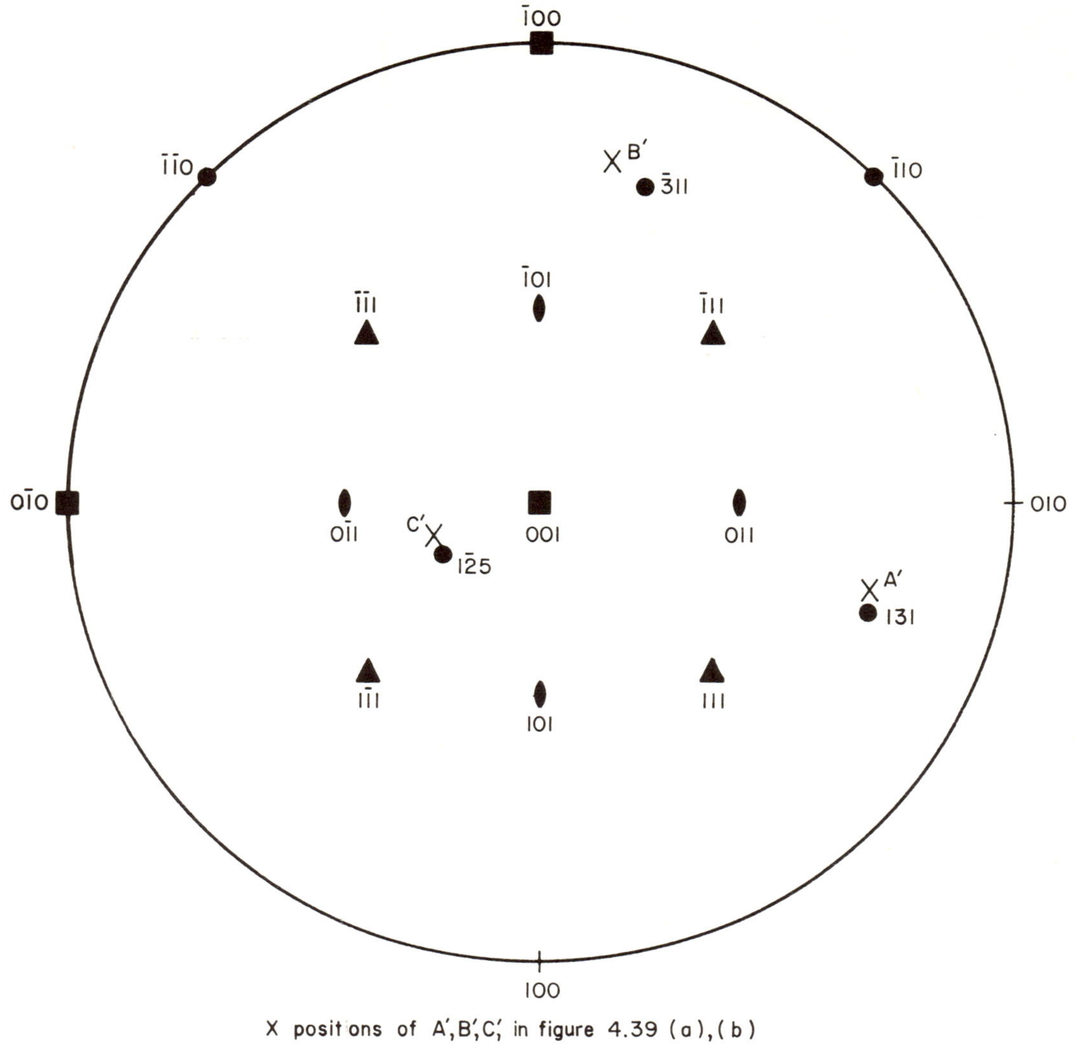

(c)
Figure 4.39 (*continued*)

with the diffraction pattern obtained in figure 4.40(a). The analysis of the diffraction pattern was completed in terms of a $(10\bar{1}2)$ twin, as shown in figure 4.40(d). Note the extra reflections produced by double diffraction.

A second example is shown in figure 4.41(a). The parent great circle is shown in figure 4.41(b), together with the trace of the twin boundary normal. Again two twins could produce the twin boundary trace, but only the $(01\bar{1}2)$ twin produces reflections consistent with the pattern in figure 4.41(a) and the indexed pattern is given in figure 4.41(c). It can be seen in figure 4.41(b) that the poles do not all lie on the same great circle although their planes give rise to diffracted beams. Reciprocal lattice streaks are responsible for this effect. Consequently, twin and parent reflections frequently occur together, although rational zone axes are seldom coincident in twin and parent and the planes are not at the exact Bragg reflecting position. Note the $\{11\bar{2}1\}$ forbidden reflections in figure 4.41(a) produced by double diffraction from two allowed reflections, that is $(0\bar{1}10)+(1\bar{1}01) = (1\bar{2}11)$.

4.13.2 *Identification of the ω Phase in a Ti–13 wt% Cr Alloy* (*after J. P. Hobday*)

In titanium and zirconium the high-temperature β phase may be stabilised by alloying so that it is retained on quenching from a high temperature, for example Zr–Nb, Ti–Nb, Ti–V. However, a fine dispersion of a metastable phase ω often forms during the quench or on subsequent ageing. A number of workers have shown that the β–ω phase mixture exhibits very characteristic diffraction patterns that enable it to be readily identified by direct comparison with published patterns. In general, the ω phase may be identified from a single SADP (such as figure 4.42(a) for which

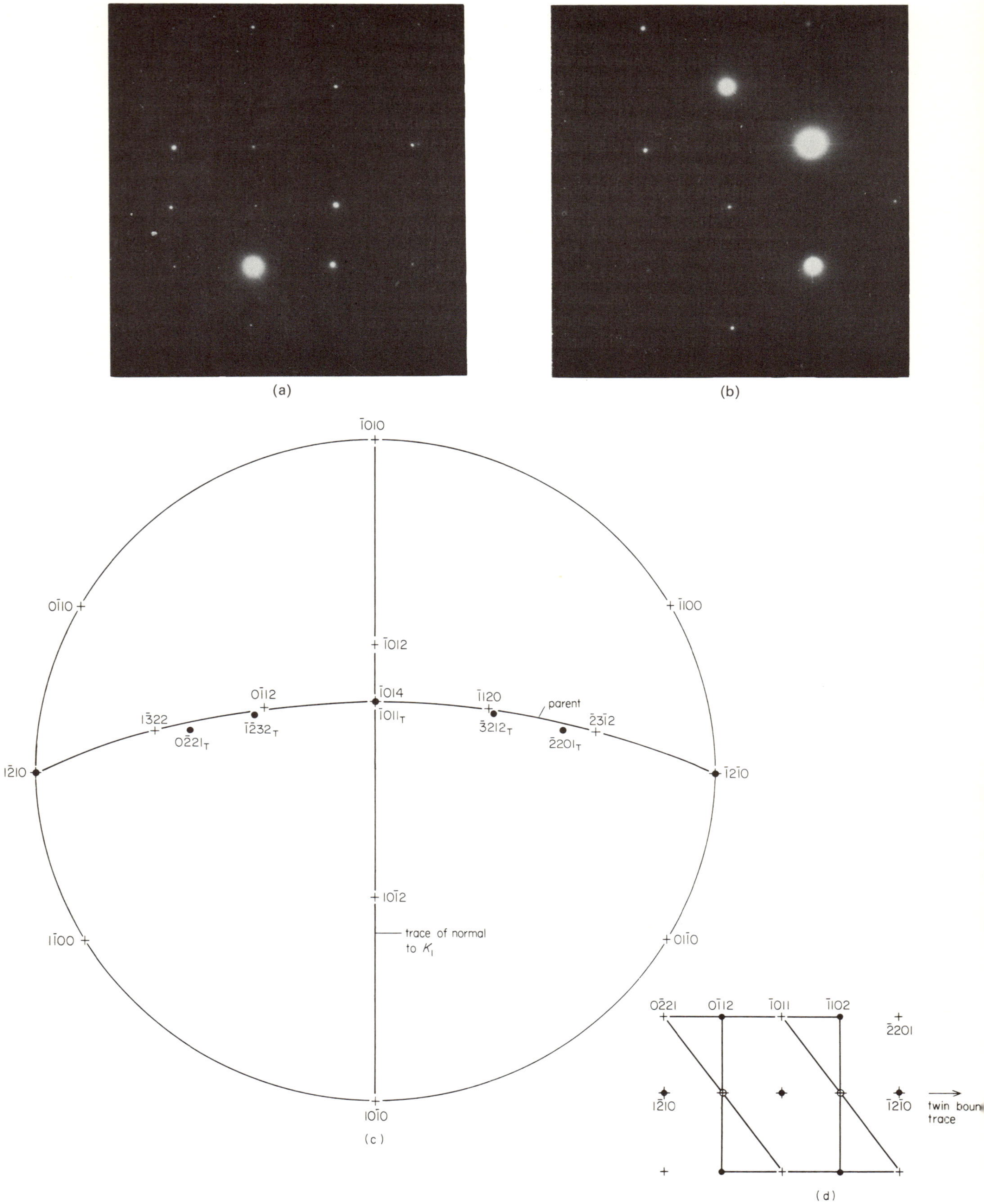

Figure 4.40 SADP from (a) twinned titanium, (b) matrix. (c) Stereographic representation of matrix pattern, and trace of twin boundary normal from (a), on standard (0001) stereogram. Twinned poles transferred from figure 2.48(a). (d) Indexed pattern [From Partridge (1967), courtesy of *Metallurgical Reviews*]

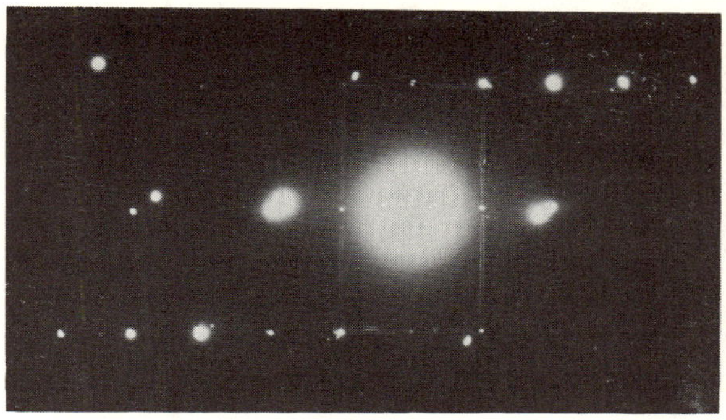

(a)

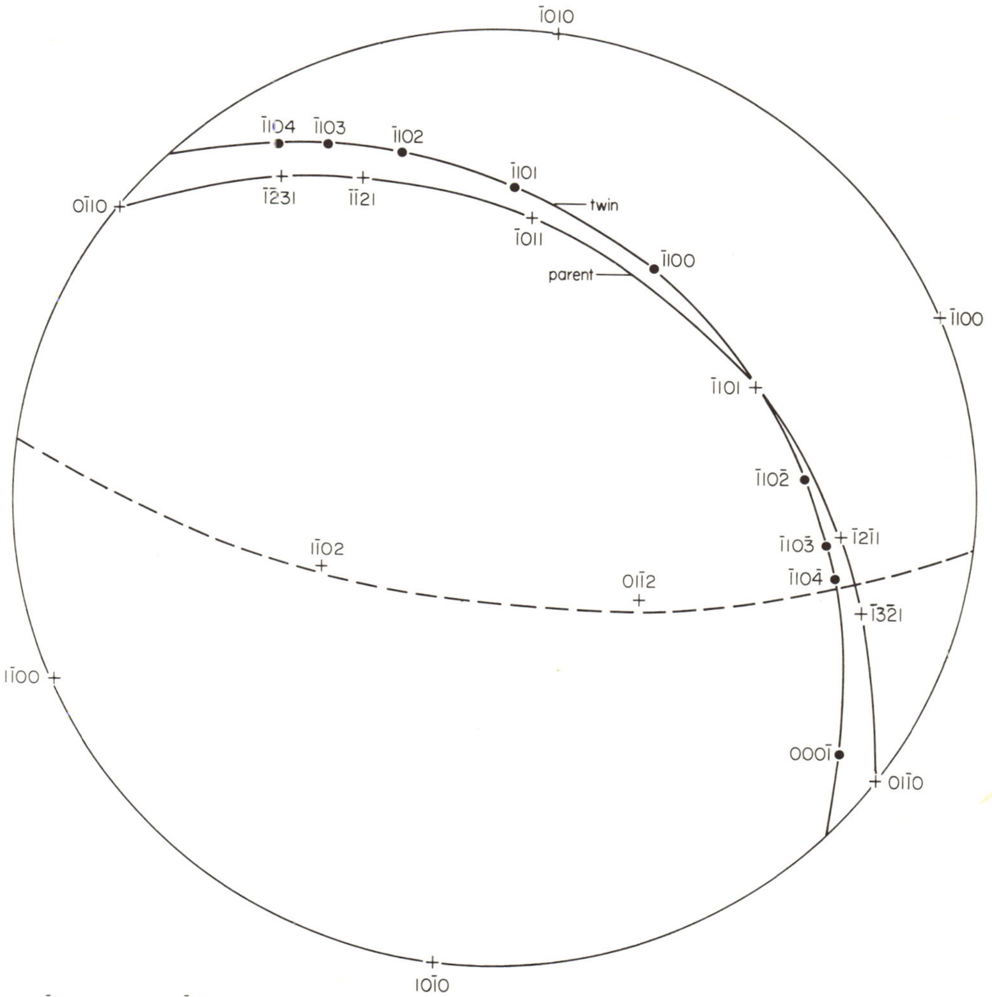

(b)

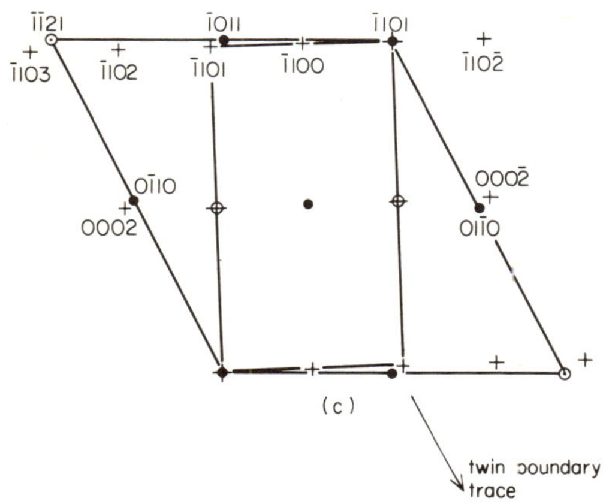

(c)

Figure 4.41 (a) SADP from twin and matrix in titanium. (b) Projection of matrix diffraction pattern, and trace of twin boundary normal from (a), on standard (0001) stereogram. Twinned poles transferred from $[01\bar{1}2]$ twin pole projection. (c) Indexed pattern, forbidden reflections produced by double diffraction [From Partridge (1967), courtesy of *Metallurgical Reviews*]

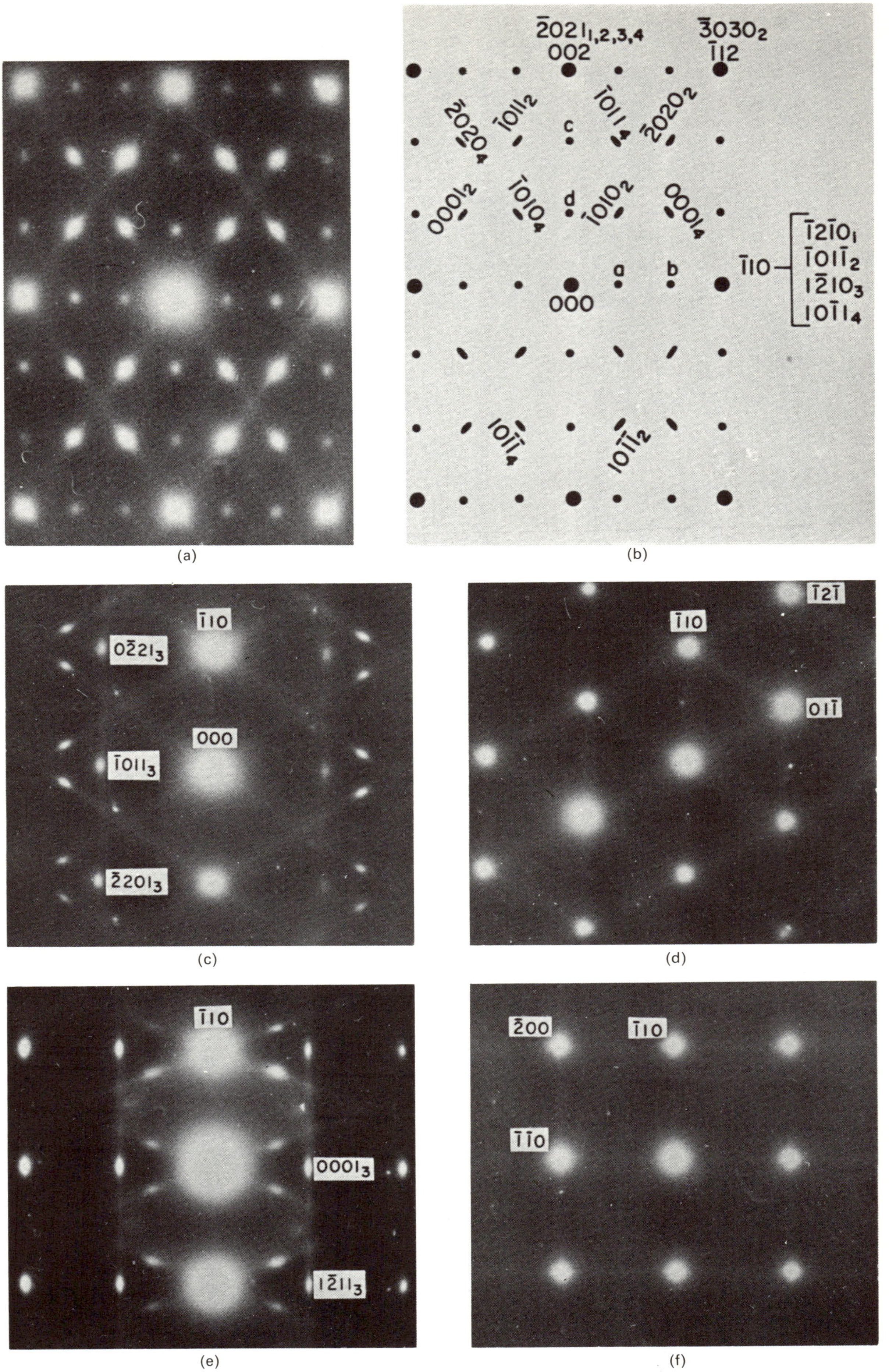

Figure 4.42 (a) $\boldsymbol{B}(110)_\beta$ diffraction pattern. (b) Schematic diffraction pattern giving indices of reflections in (a). The subscripts refer to a particular orientation of the ω phase. Diffraction patterns from the crystal tilted about a [110] axis. (c), (d) are from a different foil than (e). (f): (c) $\boldsymbol{B} \sim (221)_\beta$; (d) $\boldsymbol{B}(111)_\beta$; (e) $\boldsymbol{B} \sim (112)_\beta$; (f) $\boldsymbol{B}(001)_\beta$ [From Sass (1971), courtesy of *Acta Metallurgica*]

$\boldsymbol{B} = [110]_{\text{matrix}}$) because of its unique form. However, for the reasons outlined in section 2.13.1 complete identification should include comparison with standard patterns for other matrix $\boldsymbol{B}$. Typical patterns are shown in figure 4.42(c)–(f) for $\boldsymbol{B}[2\bar{2}1]$, $[111]$, $[112]$ and $[001]$.

The structure of the ω phase is hexagonal with interior atoms at $(\frac{1}{3}, \frac{1}{3}, \mu)$. The value of μ and c/a depends upon ageing time but for the well-developed state $\mu = \frac{1}{2}$ and $c/a = 0.613$ (Silcock *et al.*, 1956). The orientation relationship between ω and β is $(0001)_\omega//(111)_\beta$ and $[2\bar{1}\bar{1}0]_\omega//[1\bar{1}0]_\beta$; thus there are four variants of the orientation relationship corresponding to particles lying on all $\{111\}_\beta$. Using this information the SADP with $\boldsymbol{B}[110]_\beta$ may be indexed as shown in figure 4.42(b), see section 2.19.2.2. Here the fundamental β reflections are shown as filled-in circles, the reflections as circles and the double diffraction spots as crosses. The subscripts refer to the variants responsible for the individual reflections. Confirmation that reflections a–d $\boldsymbol{B}[110]$ arise from double diffraction may be obtained by tilting a few degrees away from $\boldsymbol{B}[110]$ about $[\bar{1}10]$ and $[002]$. In the case of a tilt about $[\bar{1}10]$ reflections a and b disappear when the ω reflections, for example $\{\bar{1}011\}$, $\{\bar{2}020\}$ are not excited, see figure 4.42(c).

Figure 4.43(a) shows a typical SADP taken from a Ti–13 wt % Cr alloy aged at 400 °C for 1 hour. The pattern is almost identical with that in figure 4.42(a) and clearly shows that a β–ω phase mixture is present. Figure 4.43(b) is a BF image of the area from which the SADP in figure 4.43(a) was obtained, while figure 4.43(c) is a CDF of this area obtained with the circled ω reflection, figure 4.43(a), which enables one variant of ω to be observed.

During the very early stages of ω formation, for example in the as-quenched material, considerable streaking of SADP between ω reflections

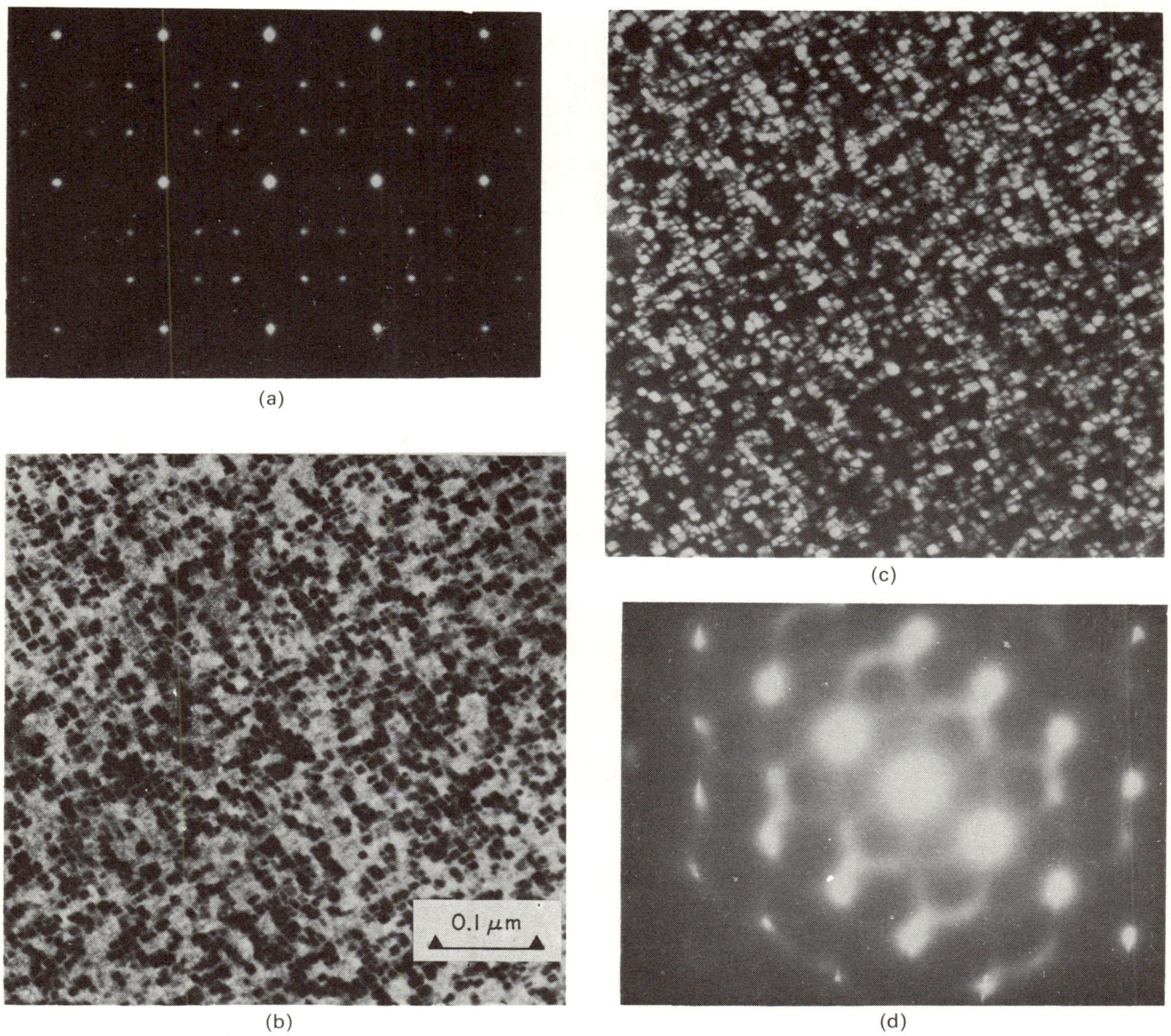

Figure 4.43 Ti–13 wt % Cr aged at 400 °C for 1 hour. (a) SADP $\boldsymbol{B}\langle 110\rangle_{\text{matrix}}$ showing ω spots. (b) BF and (c) CDF using the ω reflection arrowed. (d) Typical diffuse scattering after only 10 min at 400 °C $\boldsymbol{B}\langle 110\rangle_{\text{matrix}}$ [Courtesy of J. P. Hobday]

is observed, see figure 4.43(d). Sass (1969, 1971) has shown that these correspond to walls of intensity on $\{111\}$ in reciprocal space which may be well defined or diffuse depending upon composition. In this particular example there are complex interactions of the streaks which are not explained. The streaks from the ω phase disappear on ageing.

Note that in some alloys, for example Ti–V, the $(10\bar{1}0)$ and $(20\bar{2}0)$ superlattice reflections disappear (McCabe and Sass, 1971) implying that ω is not ordered. However, this is not true in all cases, see Balcerzack and Sass (1972).

4.13.3 *Hydrogen Contamination of Titanium Alloys (after J. D. Boyd)*

It has been established by many workers that thin foils of the hexagonal phase in titanium and its alloys can easily become contaminated with

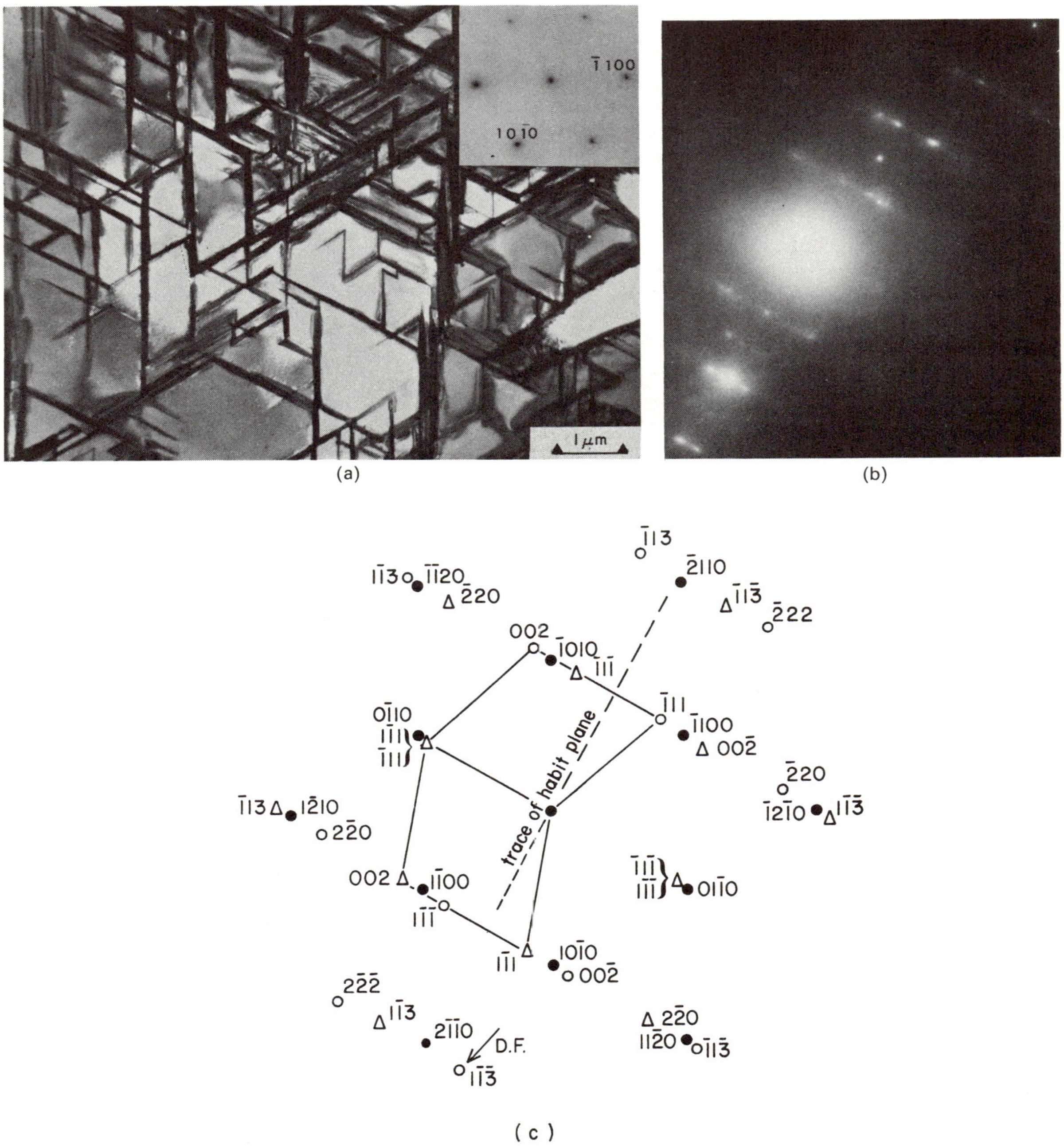

Figure 4.44 Ti–8 wt % Al–1 wt % Mo–1 wt % V charged with 800 ppm H. (a) BF image showing TiH_2 plate precipitates on $\{1\bar{1}00\}$. No diffraction spots from hydride visible (inset pattern rotated to allow for image–diffraction pattern rotation). (b) SADP including TiH_2 spots indexed in (c) with (d) a stereogram illustrating the orientation relationship $(0001)_\alpha//(110)_{TiH_2}$, $[1\bar{2}10]_\alpha//[1\bar{1}0]_{TiH_2}$ [From Boyd (1969), courtesy of *Quarterly Transactions of the American Society for Metals*]

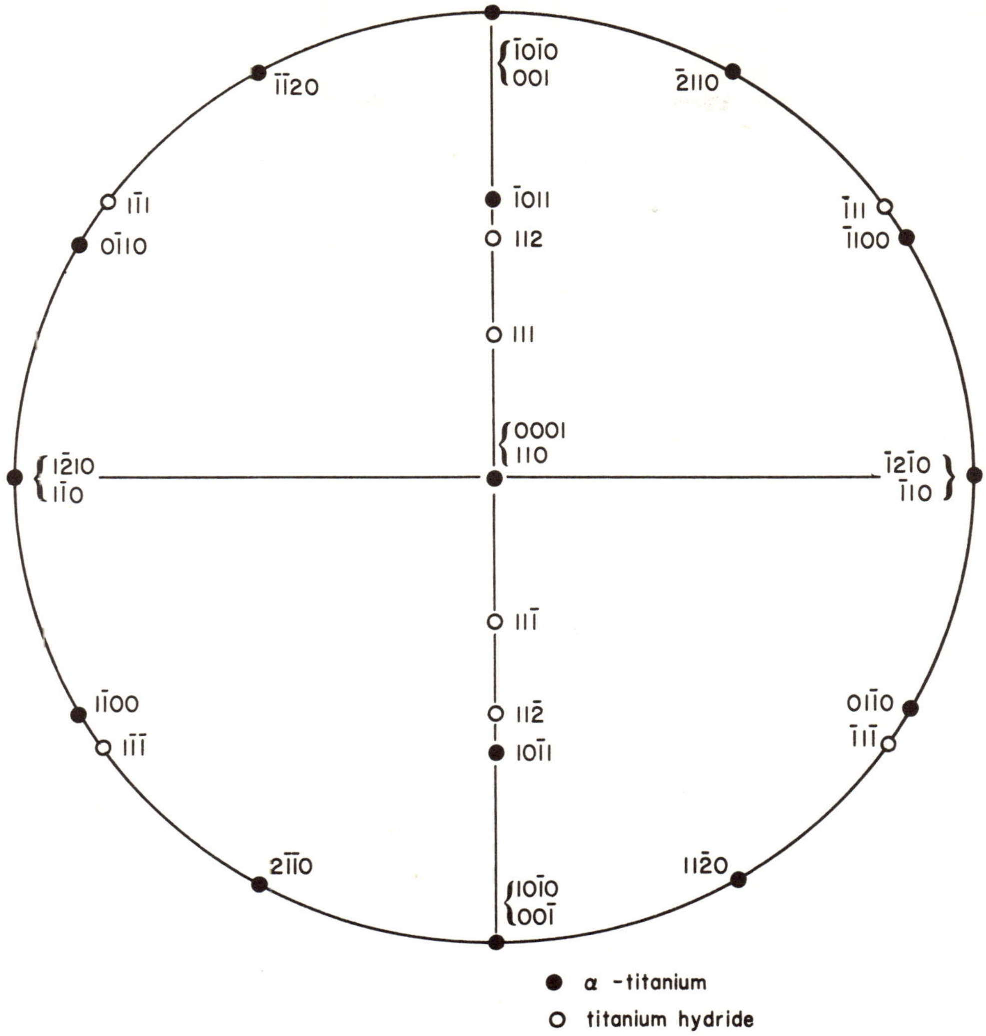

$(0001)_\alpha \parallel (110)_\gamma$, $[1\bar{2}10]_\alpha \parallel [1\bar{1}0]_\gamma$

(d)

Figure 4.44 (*continued*)

hydrogen during chemical or electropolishing (Blackburn and Williams, 1967; Sanderson and Scully, 1967; Boyd, 1969; Jones and Edington, 1972a). This manifests itself either by precipitation of hydride or by the presence of δ boundaries arising from hydrogen ordering.

Precipitation of the hydride TiH_2 is the most common occurrence. This phase is tetragonal at room temperature but the SADP obtained in the electron microscope under normal illumination conditions may be indexed (Boyd, 1969) on the bases of an f.c.c. unit cell with lattice parameters 4.45 Å ($TiH_{1.99}$) reported by Yakel (1958). Thus the indexed patterns in appendix 4 apply to this material.

Figure 4.44(a) shows typical hydride precipitates in a Ti–8 wt % Al–1 wt % Mo–1 wt % V commercial alloy containing hydrogen introduced in a controlled manner using Sievert's apparatus (Boyd, 1969). The diffraction pattern is shown inset and corresponds to $\boldsymbol{B}$ close to $[0001]_\alpha$. Trace analysis, section 2.14.1, shows that the hydride plates are on $\{10\bar{1}0\}_\alpha$ planes and intersect the surface parallel to $\langle 11\bar{2}0\rangle$, that is perpendicular to $\boldsymbol{g}\,\langle\bar{1}100\rangle$. The $\{10\bar{1}1\}$ habit plane has also been reported by Sanderson and Scully (1967). Three orientation relationships have been reported, namely

$\{0001\}_\alpha//\{001\}_H$, $\langle 1\bar{2}10\rangle_\alpha//\langle 1\bar{1}0\rangle_H$ orientation 1
$\{0001\}_\alpha//\{112\}_H$, $\langle 1\bar{2}10\rangle_\alpha//\langle 1\bar{1}0\rangle_H$ orientation 2
$\{0001\}_\alpha//\{110\}_H$, $\langle 1\bar{2}10\rangle_\alpha//\langle 1\bar{1}0\rangle_H$ orientation 3

where α refers to the hexagonal titanium matrix and the H to the hydride. A SADP illustrating orientation relationship 3 is shown in figure 4.44(b) and has been indexed, figure 4.44(c), using the rules in section 2.7.2.1, to give the orientation relationship described by the stereogram in figure

4.44(d). See section 2.11.1 for this method of describing orientation relationships.

It is important to realise that Boyd (1970) has demonstrated that hydride nucleation is assisted by deformation and is therefore likely to be more of a problem in preparing specimens from deformed material. Figure 4.45(a) shows a BF micrograph of a deformed region of the hexagonal phase in a Ti–8 wt % Al–1 wt % Mo–1 wt % V alloy while figure 4.45(b) shows the CDF micrograph obtained with the $(0\bar{1}1)$ hydride reflection, see inset. Nucleation of the hydride precipitates has occurred on the $\{10\bar{1}0\}$ slip planes.

Ordering of impurities, probably hydrogen, arising from contamination during electropolishing of Ti_3Sn has been reported by Jones and Edington (1972a). In this case, the hydrogen orders in the lattice and produces a small distortion that gives rise to δ fringes, see figure 3.37(b),(c). This effect can be recognised from the characteristic δ fringe contrast produced (asymmetric BF fringe image, symmetric DF image, sections 3.8.1, compare figure 3.35(a),(b)).

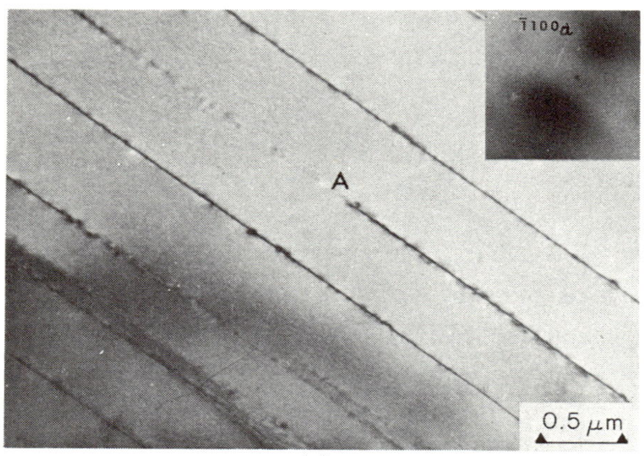

(a)

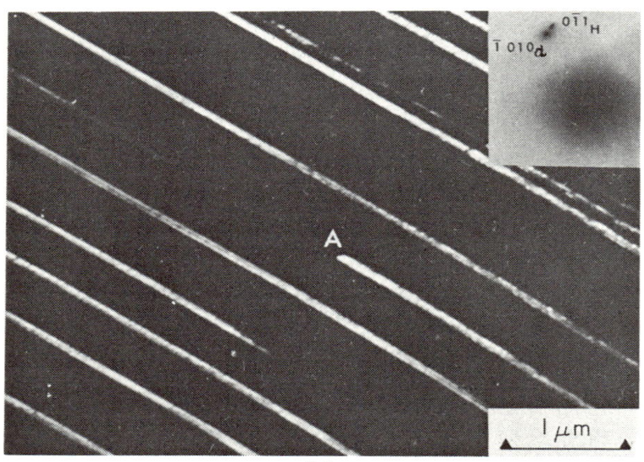

(b)

Figure 4.45 Ti–8 wt % Al–1 wt % Mo–1 wt % V charged with 400 ppm H then deformed ~2% at 20 °C. Strain-induced hydride plates on $(\bar{1}100)$ are visible in the BF image (a), and the CDF (b) obtained with $(0\bar{1}1)$ TiH_2 reflection [From Boyd (1969), courtesy of *Quarterly Transactions of the American Society for Metals*]

Note that significant hydrogen contamination can be avoided using the electropolishing solution suggested by Blackburn and Williams (1967).

4.13.4 *Determination of the Dislocation Burgers Vectors in the Intermetallic Compound Ti_3Sn (after P. J. Jones)*

This compound has a hexagonal DO_{19} crystal structure and exhibits only limited ductility at low temperatures (Jones and Edington, 1973). After small strains in the temperature range 196–400 °C slip bands containing dislocations may be observed in thin foils and these have been analysed using the computer image matching technique first developed by Head (1967) and modified by Humble (1968), see section 3.4.3.2. The computer program used in the work enabled undissociated dislocations to be treated in hexagonal crystal structures while allowing for elastic anisotropy (Jones and Edington, 1972a,b). The extinction distances were calculated with the method described in section 3.2.3.1, using the electron scattering factors in appendix 8. The deviation $\boldsymbol{s}$ from the exact Bragg reflecting position was determined using the method outlined in section 2.10.2. Since the elastic constants of Ti_3Sn are unknown, the pure titanium elastic of Yoo and Wei (1967) were used to calculate the anisotropy parameters

$$A_{\text{shear}} = \frac{2C_{44}}{C_{11}-C_{12}} = 1.33$$

and

$$A_{\text{compression}} = \frac{C_{11}+C_{12}-2C_{13}}{C_{33}-C_{13}} = 1.04$$

where C_{11}, C_{12}, C_{13}, C_{33} and C_{44} are the five independent elastic stiffness constants. Jones and Edington (1972b) pointed out that the main effect of elastic anisotropy is to lower the value of $\boldsymbol{g}\cdot\boldsymbol{b}\wedge\boldsymbol{u}$ at which images becomes visible when $\boldsymbol{g}\cdot\boldsymbol{b} = 0$. Since titanium is almost elastically isotropic, this point occurs close to $\boldsymbol{g}\cdot\boldsymbol{b}\wedge\boldsymbol{u} = 0.64$ and this was also found to be true for Ti_3Sn. Consequently, the use of the titanium elastic constants is a reasonable first approximation.

In some cases examples of invisibility, or weak contrast, were not obtained for any of the operative reflections because of the large values for $\boldsymbol{g}\cdot\boldsymbol{b}\wedge\boldsymbol{u}$. However, sometimes it was possible to obtain two examples of near invisibility with different reflections, which greatly simplified the image matching procedure, because only a limited amount of computation was necessary.

As pointed out in section 3.4.3.1, the '$\boldsymbol{g}\cdot\boldsymbol{b} = 0$ effective invisibility' criterion may be used to estimate the Burgers vector before computation. Such an example is discussed below.

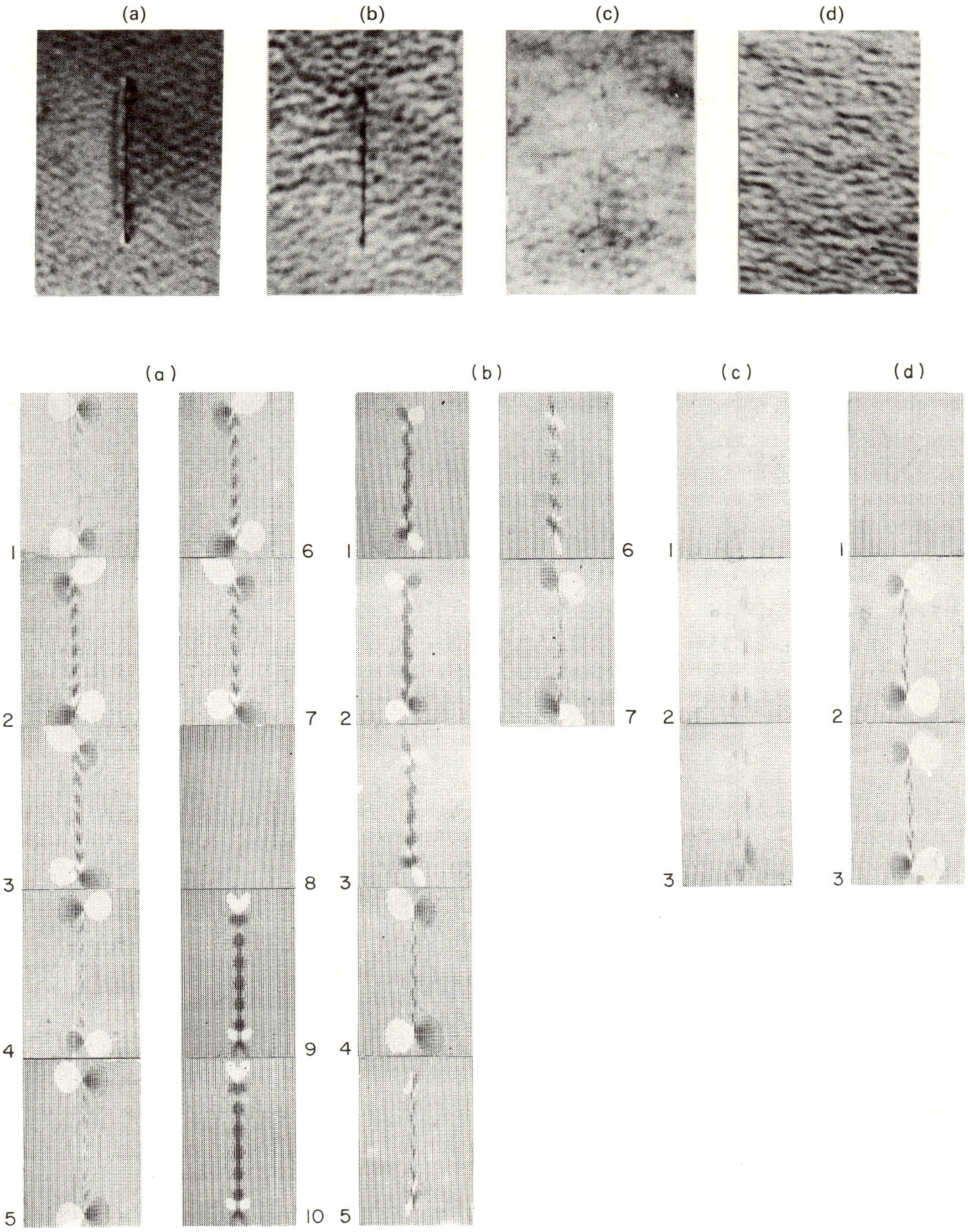

Figure 4.46 BF images of a dislocation in magnesium. A dislocation lying along $[2\bar{1}\bar{1}0]$ and foil normal $[11\bar{2}0]$ observed at 100 kV. $\boldsymbol{B}$ is the beam direction. (a) = $\boldsymbol{g}(0002)$, $\boldsymbol{B}[11\bar{2}0]$, $w = 0.10$; (b) = $\boldsymbol{g}(2\bar{2}01)$, $\boldsymbol{B}[11\bar{2}0]$, $w = 0.20$; (c) = $\boldsymbol{g}(\bar{4}400)$, $\boldsymbol{B}[11\bar{2}0]$, $w = 1.5$; (d) = $\boldsymbol{g}(\bar{4}220)$, $\boldsymbol{B}[01\bar{1}0]$, $w = 0.15$. (a) 1–10 Computed images for the conditions in (a) for the Burgers vectors: 1, $[0001]$; 2, $[11\bar{2}\bar{3}]$; 3, $[11\bar{2}3]$; 4, $\frac{1}{3}[2\bar{1}\bar{1}3]$; 5, $\frac{1}{3}[2\bar{1}\bar{1}0]$; 6, $\frac{1}{3}[\bar{1}2\bar{1}3]$; 7, $[\bar{1}2\bar{1}3]$; 8, $\frac{1}{3}[2\bar{1}\bar{1}0]$; 9, $\frac{1}{3}[\bar{1}2\bar{1}0]$; 10, $[11\bar{2}0]$. (b) 1–7 Computed images in (b) for the Burgers vectors: 1, $[0001]$; 2, $\frac{1}{3}[11\bar{2}\bar{3}]$; 3, $\frac{1}{3}[11\bar{2}3]$; 4, $\frac{1}{3}[2\bar{1}\bar{1}3]$; 5, $\frac{1}{3}[2\bar{1}\bar{1}0]$; 6, $\frac{1}{3}[\bar{1}2\bar{1}3]$; 7, $\frac{1}{3}[\bar{1}2\bar{1}3]$. (c) 1–3 Computed images for the conditions in (c) for the Burgers vectors: 1, $[0001]$; 2, $\frac{1}{3}[11\bar{2}\bar{3}]$; 3, $\frac{1}{3}[11\bar{2}3]$. (d) 1–3 Computed images for the conditions in (d) for the Burgers vectors: 1, $[0001]$; 2, $\frac{1}{3}[11\bar{2}\bar{3}]$; 3, $\frac{1}{3}[11\bar{2}3]$ [From Jones and Edington (1972b), courtesy of *The Philosophical Magazine*]

Figure 4.46 shows a typical case with two cases of 'effective invisibility'. The reflections concerned are (0002), $(2\bar{2}0\bar{1})$, $(\bar{4}400)$ and $(\bar{4}200)$ and the dislocation is 'effectively invisible' in the last two. The line direction of the dislocation is $[2\bar{1}\bar{1}0]$ and the foil normal is $[11\bar{2}0]$. Here the Burgers vector is most likely to be [0001], since this is the zone axis of $(\bar{4}400)$ and $(\bar{4}220)$, but nevertheless all the Burgers vectors were checked by computation for the case in figure 4.46(a), see figure 4.46(a), 1–10. Clearly $\frac{1}{3}[2\bar{1}\bar{1}0]$ can be eliminated because it would be invisible. In addition $\frac{1}{3}[\bar{1}2\bar{1}0]$ and $\frac{1}{3}[11\bar{2}0]$ can both be rejected because of the form of the image which is characteristic of $\boldsymbol{g}\cdot\boldsymbol{b} = 0$, $\boldsymbol{g}\cdot\boldsymbol{b}\wedge\boldsymbol{u} > 0.64$. Figure 4.46(b), 1–7 shows the computed images for the remaining Burgers vectors for the experimental conditions in figure 4.46(b). Comparing both sets of computed micrographs with the experimental ones it is clear that Burgers' vectors [0001], $\frac{1}{3}[11\bar{2}\bar{3}]$ and $[11\bar{2}3]$ match the experimental images reasonably well for both these reflections. However, figure 4.46(c), 1–3, and figure 4.46(d), 1–3, show that the correct Burgers vector is indeed [0001].

Only a small fraction of the dislocations identified in deformed material had Burgers' vectors [0001] and these lay in line directions contained in $\{10\bar{1}0\}$ or $\{11\bar{2}0\}$. Most of the dislocations had Burgers' vectors $\frac{1}{3}\langle 11\bar{2}0\rangle$ and lay on $\{10\bar{1}0\}$, $\{10\bar{1}1\}$ or (0001). Both Burgers' vectors were identified in networks in undeformed material but again $\frac{1}{3}\langle 11\bar{2}0\rangle$ predominated.

V. EXAMPLES OF ANALYSES IN NON-METALS

In the following sections are included a number of examples of the use of electron microscopy on non-metallic materials, including ceramics, polymers, semi-conductors and minerals.

4.14 Ceramics

Electron microscopy has been applied to a number of problems in ceramics. Studies of the defect structure have been reported (Ashbee and Smallman, 1963; Ashbee *et al.*, 1963; van Landuyt, 1966; van Landuyt *et al.*, 1964) and order/disorder phenomena investigated (Billingham *et al.*, 1972). In this section an example is included of an order/disorder study in vanadium carbide.

4.14.1 *Ordering of Vacancies in* $VC_{0.83}$ *(after R. J. Hannick)*

In common with many other MC carbides, vanadium carbide exists over a wide range of composition extending from $VC_{0.75}$–$VC_{0.9}$ below 1000 °C. However, like other MC carbides, different ordered crystal structures exist within this phase field. For reviews, see Lewis *et al.* (1972) and Billingham *et al.* (1972). The overall crystal structure of MC carbides is cubic and composition variations are accommodated by the presence of vacancies on the carbon sublattice. The different ordered structures are formed by ordering of the vacancies on this sublattice.

In the case of vanadium carbide there is a cubic superlattice near $VC_{0.87}$ (V_8C_7) with a lattice parameter twice that of the sodium chloride structure (de Novion, 1966). A more complicated situation exists in the vicinity of the composition $VC_{0.83}$ ($\sim V_6C_5$) where both hexagonal (Venables *et al.*, 1968) and monoclinic (Billingham *et al.*, 1972) superlattices have been reported.

An example is shown here of the determination of the hexagonal superlattice, together with the use of BF techniques in the detailed interpretation of images and SADP.

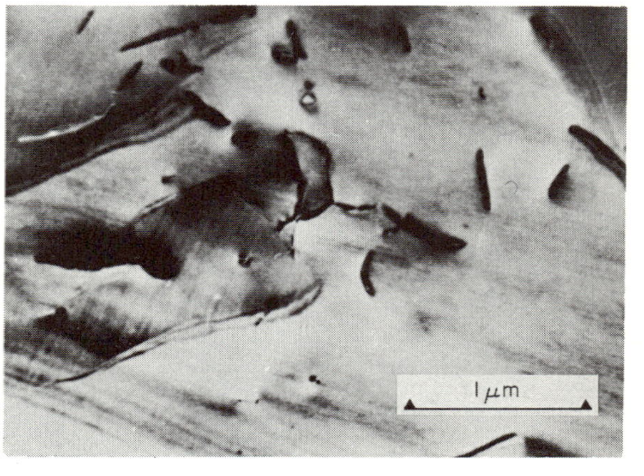

Figure 4.47 Vanadium carbide V_6C_5. (a) BF image of an axial domain containing grown-in dislocations [From Hannick *et al.* (1971), courtesy of *The Philosophical Magazine*] (b) SADP showing strong spots from the fundamental sodium chloride lattice and ordered V_6C_5 (fainter spots) $\boldsymbol{B}[111]$ sodium chloride lattice

A typical image of slowly cooled, that is strongly long-range ordered, material is shown in figure 4.47, together with the SADP. In this case the spots marked F may be indexed using the rules in section 2.7.2.1 on the basis of the fundamental sodium chloride crystal structure of VC with lattice parameter of 4.16 Å. However, the remaining less-intense reflections clearly arise from a superlattice with a larger unit cell. Thus larger interplanar spacings exist and they produce spots nearer the transmitted beam than the fundamental cubic carbide reflections. Consequently, the image in figure 4.47 may be tentatively concluded to consist of a single ordered domain surrounded by domain walls.

In order to define the symmetry of the superlattice unit cell, SADP from a number of simple ***B*** with respect to the fundamental reflections must be obtained by tilting the specimen, see section 2.10.1. Figure 4.48 shows SADP with ***B***⟨100⟩, ⟨110⟩ and ⟨111⟩ in the NaCl fundamental crystal structure. To facilitate interpretation each pattern has been obtained from an individual domain grown to a large size by slowly cooling the material from a high temperature.

Inspection of the SADP in figure 4.48 shows that each pattern has two-fold symmetry about ***B*** except that in figure 4.48(i) which has six-fold symmetry. This may be interpreted either as a six-fold symmetry axis in the hexagonal crystal class or a three-fold axis in a cubic crystal class (which in the SADP has six-fold symmetry because of the presence of positive and negative reflections, see appendix 4). If the superlattice were cubic, figure 4.48(i) would be indexed as a cube–cube orientation relationship between superlattice and fundamental NaCl reflections. Consequently, the SADP with [100] should have a four-fold symmetry, that is a cubic pattern of superlattice reflections but this is not observed and this superlattice may be rejected. We may now proceed on the basis of an hexagonal unit cell.

It now remains to assign indices to the extra spots in the SADP and to determine the lattice parameters *c* and *a*. To begin with, a list of measured interplanar spacings from the SADP is

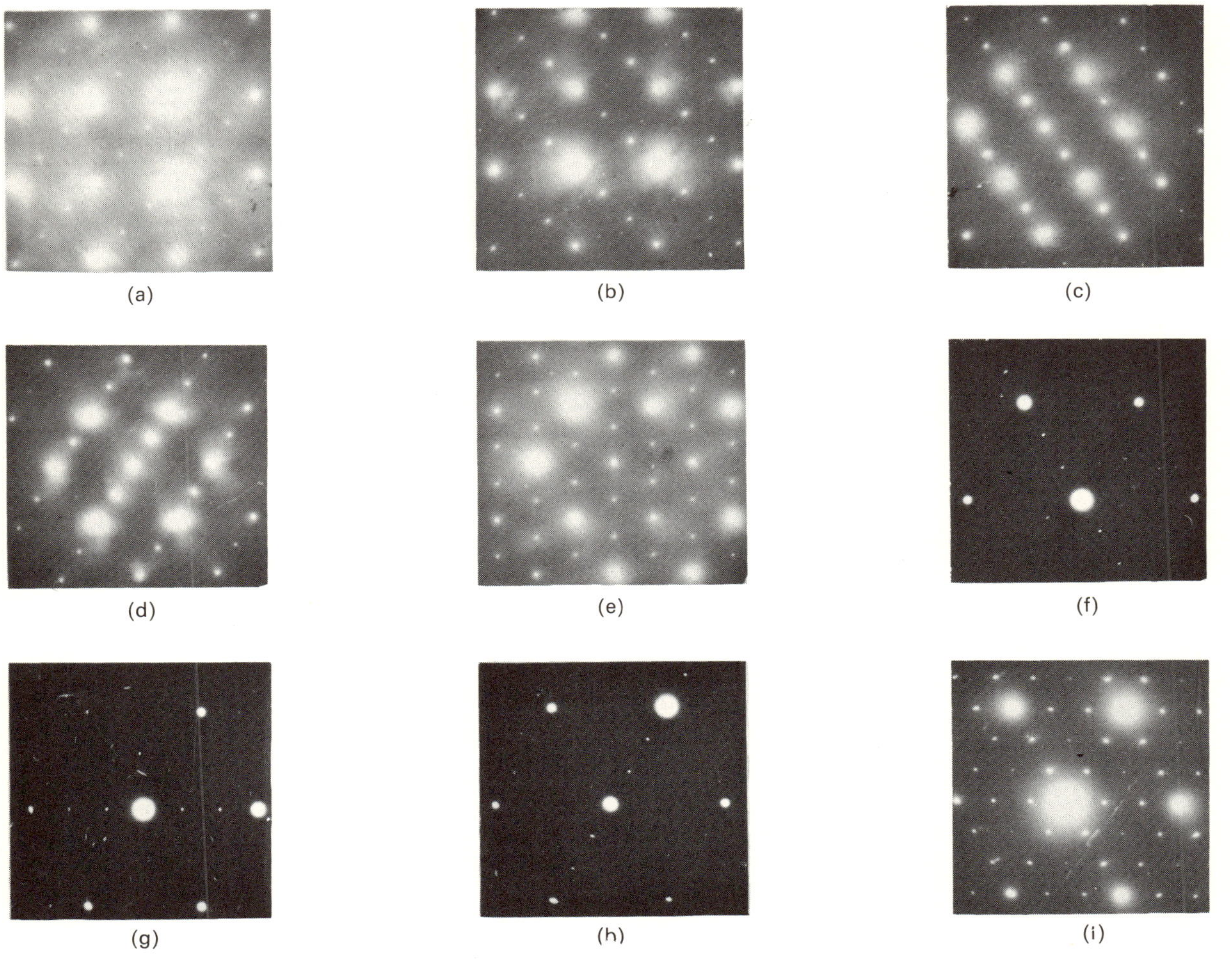

Figure 4.48 SADP taken from the V_6C_5 crystal with (a), (b) ***B***⟨100⟩ in the fundamental sodium chloride lattice; (c)–(e) ***B***⟨110⟩ in the fundamental sodium chloride lattice; (f)–(i) ***B***⟨111⟩ in the fundamental sodium chloride lattice from individual ordered domains

Table 4.23 Spacing of the various superlattice reflections as a fraction of the {220} fundamental spacing[a], together with the associated interplanar spacings

Beam direction fundamental	Spot spacing (relative to {220} fundamental)	Direction in fundamental superlattice	Interplanar spacing (Å)
111	(1) 0.333	220	4.412
	(2) 0.572		2.567
	(3) 1.221		1.204
	(4) 0.844		1.224
110	(5) 0.333	220	4.412
	(6) 0.710	200	2.068
	(7) 0.997	111	4.788
	(8) 0.393		3.738
	(9) 0.972		1.511
	(10) 1.075		1.366
	(11) 0.910		1.614
100	(12) 0.333	220	4.410
	(13) 1.000		1.470
	(14) 0.713		2.060
	(15) 0.526		2.791
	(16) 1.119		1.231

[a] Interplanar spacing {220} = 1.47 Å (a = 4.157 Å).

made. This has been done in table 4.23 in which the measurements on the plate have been expressed as a ratio of the distance from the {220} spot to the centre spot. Using the six-fold symmetry of figure 4.48(i) (that is ***B*** must be [0001] superlattice) we may index as a trial the closest spot as $\{10\bar{1}0\}$ the largest interplanar spacing in the basal plane. Then the lattice parameter may be calculated using the equation in table A1.1, appendix 1, which gives 5.09 Å. Since the c direction lies parallel to the ⟨111⟩ fundamental directions, it is now necessary to obtain the d spacing of the closest spot in ⟨111⟩ which is 4.788 Å for ***B***⟨110⟩ (see table 4.23). This spot may be indexed as (0001) (0002) (0003), etc., depending upon the structure factor. It is now necessary to work out interplanar spacings and interplanar angles for those reflections for which $l \neq 0$, that is which depend on the ratio of c/a. These may be cross-checked against the SADP ***B***⟨100⟩ and ⟨110⟩ to obtain consistent indexing of the patterns (***B***⟨111⟩

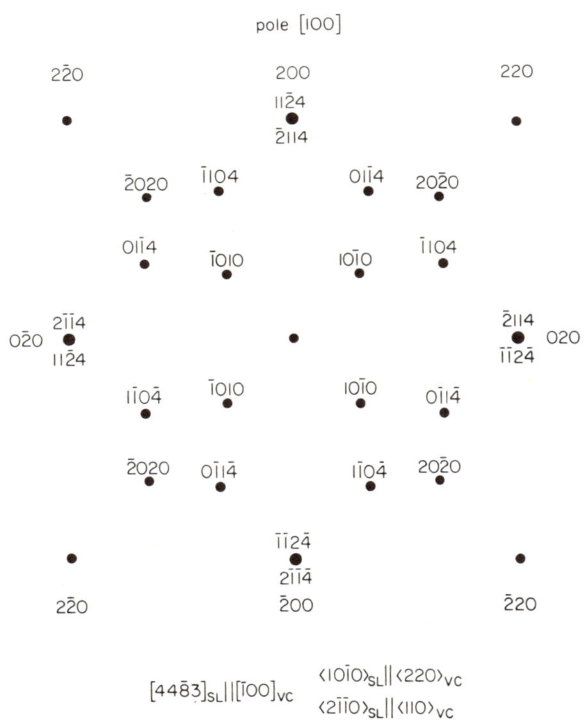

(a)

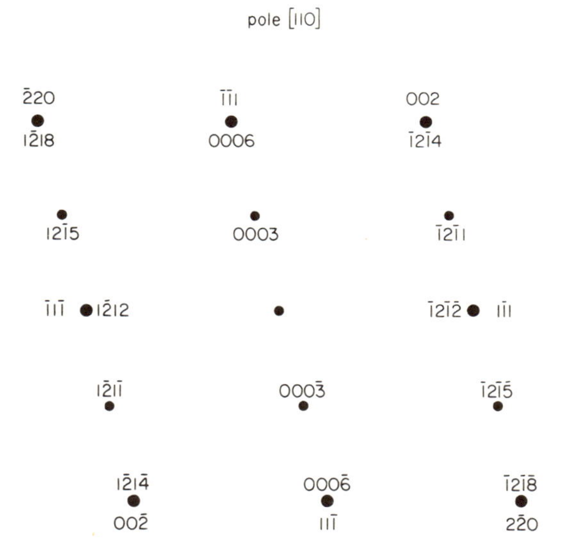

(b)

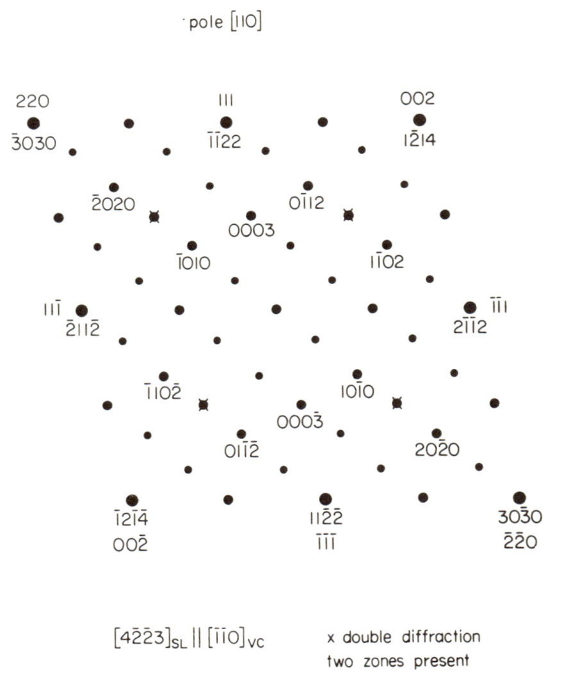

(c)

Figure 4.49 Indexed SADP corresponding to figure 4.48: (a) ***B***[100], (b), (c) ***B***[110], (d), (e) ***B***[111]

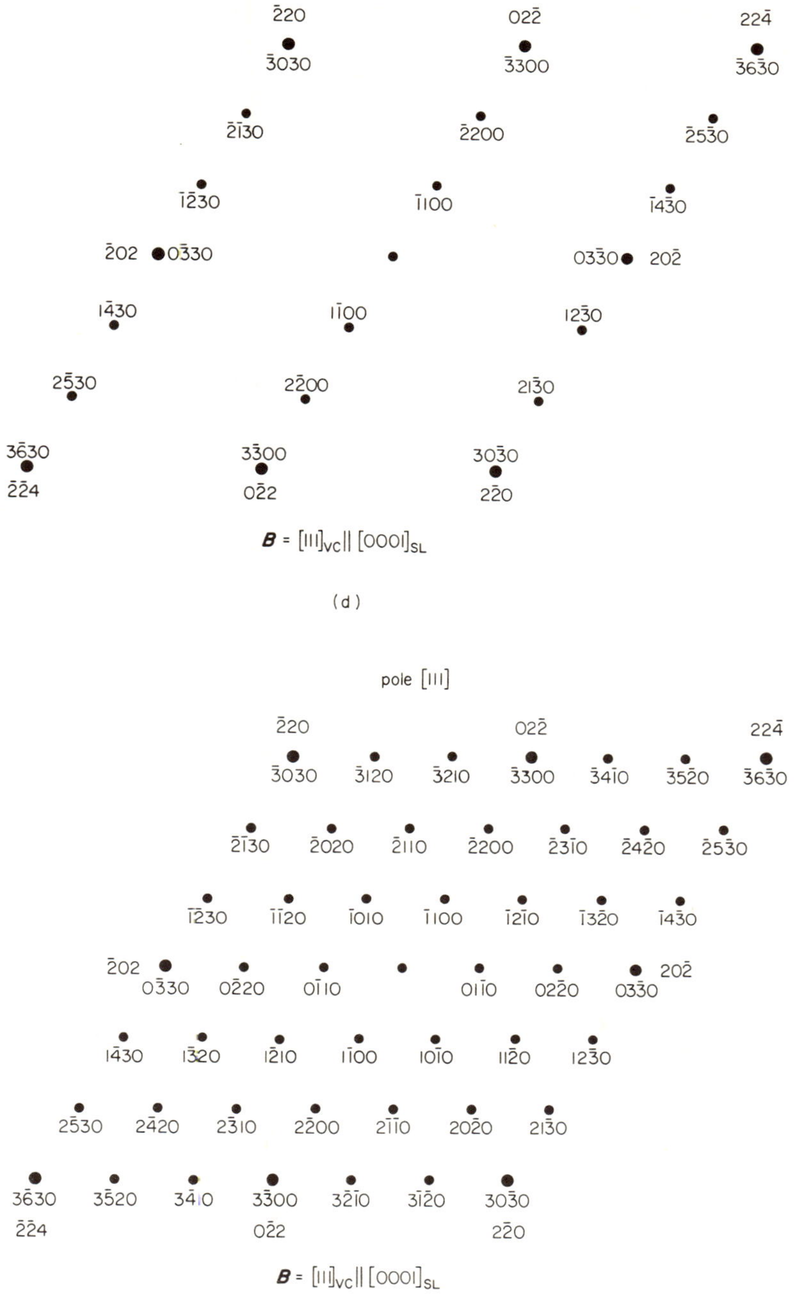

Figure 4.49 (*continued*)

contains only reflections perpendicular to the basal plane, that is $l = 0$). In fact this reflection should be indexed as (0003), that is $c = 14.2$ Å, and the interplanar spacings are those shown in table 4.24.

It is necessary to appreciate that, since $[0001]_{\text{superlattice}} // \langle 111 \rangle_{\text{fundamental NaCl}}$, four types of domain can exist with the c axis parallel to each $\langle 111 \rangle$. Thus the patterns in figure 4.48 correspond to different types (variants) of orientation of superlattice. For clarity, figure 4.49(a) includes indexing from both figure 4.48(a) and figure 4.48(b). Figure 4.49(b) and figure 4.49(c) correspond to two of the four variants with $\boldsymbol{B}[110]$ fundamental and similarly for figure 4.49(d) and figure 4.49(e) with $\boldsymbol{B}[111]$.

Note that, apart from the crystal structure and lattice parameter of the unit cell, no further information can be obtained without recourse to other techniques. Venables *et al.* (1968) have shown, using nuclear magnetic resonance, that the super-

Table 4.24 Interplanar spacings for VC and V_6C_5 (a = 5.09 Å, c = 14.2 Å)

VC		V_6C_5	
d values calculated from cubic unit cell a = 4.157 Å		*d* values calculated from hexagonal unit cell a = 5.09 Å, c = 14.2 Å	
hkl	*d* (Å)	*hkil*	*d* (Å)
		0003	4.8
		$10\bar{1}0$	4.41
		$10\bar{1}2$	3.76
		$10\bar{1}4$	2.79
		$11\bar{2}0$	2.54
		$11\bar{2}1$	2.51
111	2.40	0006 } $\bar{1}\bar{1}22$ }	2.40
		$20\bar{2}0$	2.20
		$10\bar{1}6$	2.11
200	2.08	$11\bar{2}4$	2.08
		$11\bar{2}5$	1.007
		$20\bar{2}4$	1.88
		$12\bar{3}0$	1.67
		$20\bar{2}6$	1.62
		0009	1.60
220	1.47	$30\bar{3}0$ } $11\bar{2}8$ }	1.47
		$\bar{2}\bar{2}41$	1.27
113	1.25	$30\bar{3}6$	1.25
		$13\bar{4}0$	1.22
222	1.20	$\bar{2}\bar{2}44$	1.20
		$22\bar{4}5$	1.16
		$30\bar{3}9$	1.08
400	1.04	$22\bar{4}8$	1.04
		$23\bar{5}0$	1.01
224	0.85	$33\bar{6}0$	0.85

lattice arises from ordering of the carbon vacancies in the sheets of carbon atoms lying parallel to the {111} of the fundamental carbide lattice.

4.15 Polymers

Polymers may be either 'crystalline' or 'amorphous'. The former have received more study than the latter using electron microscopy but examples of typical investigations from both types of material are included here.

4.15.1 *Stress-induced Twinning in Polyethylene Single Crystals (after P. Allan)*

Stress-induced twinning and phase transformations have been reported observed in both bulk and single crystals of polyethylene. For a review of the observations on compressed, drawn or annealed and drawn material, see Hay and Keller (1970). In this section the determination of twinning modes in single-crystal polyethylene using transmission electron microscopy is described. Twinning elements K_1, K_2, η_1, η_2 are described in section 2.12. Using this approach the crystallographic data on the twinning modes may be established using SADP.

Polyethylene single crystals consist of diamond-shaped sheets approximately 100–200 Å thick and may be observed for a limited period of time in the electron microscope without degradation provided low levels of illumination are used (Grubb and Groves, 1971; Grubb *et al.*, 1972a,b). These single-crystal specimens may be deformed on a Mylar substrate using the procedure described by Kiho *et al.* (1964), then stripped from the substrate and deposited on a carbon replica for viewing in the electron microscope (Allan *et al.*, 1973). Single crystals of polyethylene can have an orthorhombic or monoclinic crystal structure which may coexist together so that SADP can contain spots from both forms. However, the example described here only involves the orthorhombic form, see table 4.25 for lattice parameters and interplanar spacings.

Figure 4.50(a) shows a typical deformed crystal ~200 Å thick, while figure 4.50(b) is its diffraction pattern. Using the interplanar spacings in table 4.25 and the interplanar angles calculated from the equation in table A1.1, the SADP in figure 4.50(b) may be indexed using the procedure outlined in section 2.7.2.1. In fact, it consists of four patterns shown schematically in figure 4.50(c); each symbol describes the position of spots from one pattern. Figure 4.50(d) shows one indexed pattern consisting of large filled-in triangles; the open triangles correspond to missing reflections. ***B*** is $[00\bar{1}]$. The remaining patterns may be indexed in the same way and all have ***B*** ⟨001⟩.

In figure 4.50(c), the various patterns are twin-related. Thus the patterns consisting of large triangles or small circles superimpose on a 180° rotation about XY, figure 4.50(e). This situation corresponds to viewing the twin in a direction perpendicular to η_1 and in K_1, figure 2.43(c). Thus, $K_1 = \{100\}$ $\eta_1 = \langle 110\rangle$. The same relationship occurs between the small triangles and the large circles in figure 4.50(c). Furthermore, by a similar analysis both parts of twin-related patterns are in a K_1 {310} η_1 ⟨310⟩ relationship with one another.

Table 4.25 Interplanar spacings for polyethylene (orthorhombic) crystal structure; a = 7.4 Å, b = 4.93 Å, c = 2.534 Å (Bunn, 1939)

Plane indices	*d* spacing (Å)	Plane indices	*d* spacing (Å)
110	4.102	320,410,121	1.729
200	3.696	113	1.663
210	2.956	130,221	1.596
020	2.467	230	1.502
120	2.340	510,321,411	1.435
011	2.254	330,031	1.379
310	2.203	231	1.292
111	2.156	520,002	1.267
201	2.089	040,600,511	1.236
220	2.050	112	1.210
211	1.924	022	1.127
400	1.848	530,312	1.098

Bevis and Crellin (1971) have analysed possible twin modes in polyethylene and have obtained a range of possible combinations of K_1, K_2, η_1, η_2 with a shear 1.5. These are shown in table 4.26 and it is clear that the SADP in figure 4.50 corresponds to twinning of types 1 and 2.

4.15.2 *Order in Amorphous Polystyrenes* (*after G. S. Y. Yeh*)

It has been generally accepted that amorphous polymers are an orderless and structureless mass of coiled and intercoiled molecules. However, results from structural studies including radial

Table 4.26 Possible two-dimensional twinning shear modes with the shear 1.33 (Bevis and Crellin, 1971)

	K_1	K_2	η_1	η_2	Shear
1	310	$1\bar{1}0$	$1\bar{3}0$	110	0.25
2	$1\bar{1}0$	310	110	$1\bar{3}0$	0.25
3	120	100	210	010	0.67
4	100	120	010	210	0.67
5	110	110	110	$\bar{1}10$	0.83
6	110	110	$\bar{1}10$	110	0.83
7	$\bar{1}10$	1,1.26,0	110	$\overline{1.26}$,1,0	1.08
8	1,1.26,0	$\bar{1}10$	$\overline{1.26}$,1,0	110	1.08
9	1,4.2,0	3.37,1,0	$\overline{4.2}$,1,0	$\bar{1}$,3.37,0	1.3
10	3.37,1,0	1,4.2,0	$\bar{1}$,3.37,0	$\overline{4.2}$,1,0	1.3
11	110	100	$1\bar{1}0$	010	1.33
12	100	110	010	$1\bar{1}0$	1.33

Clearly twins 1 and 2 satisfy the twin patterns in figure 4.50(b).

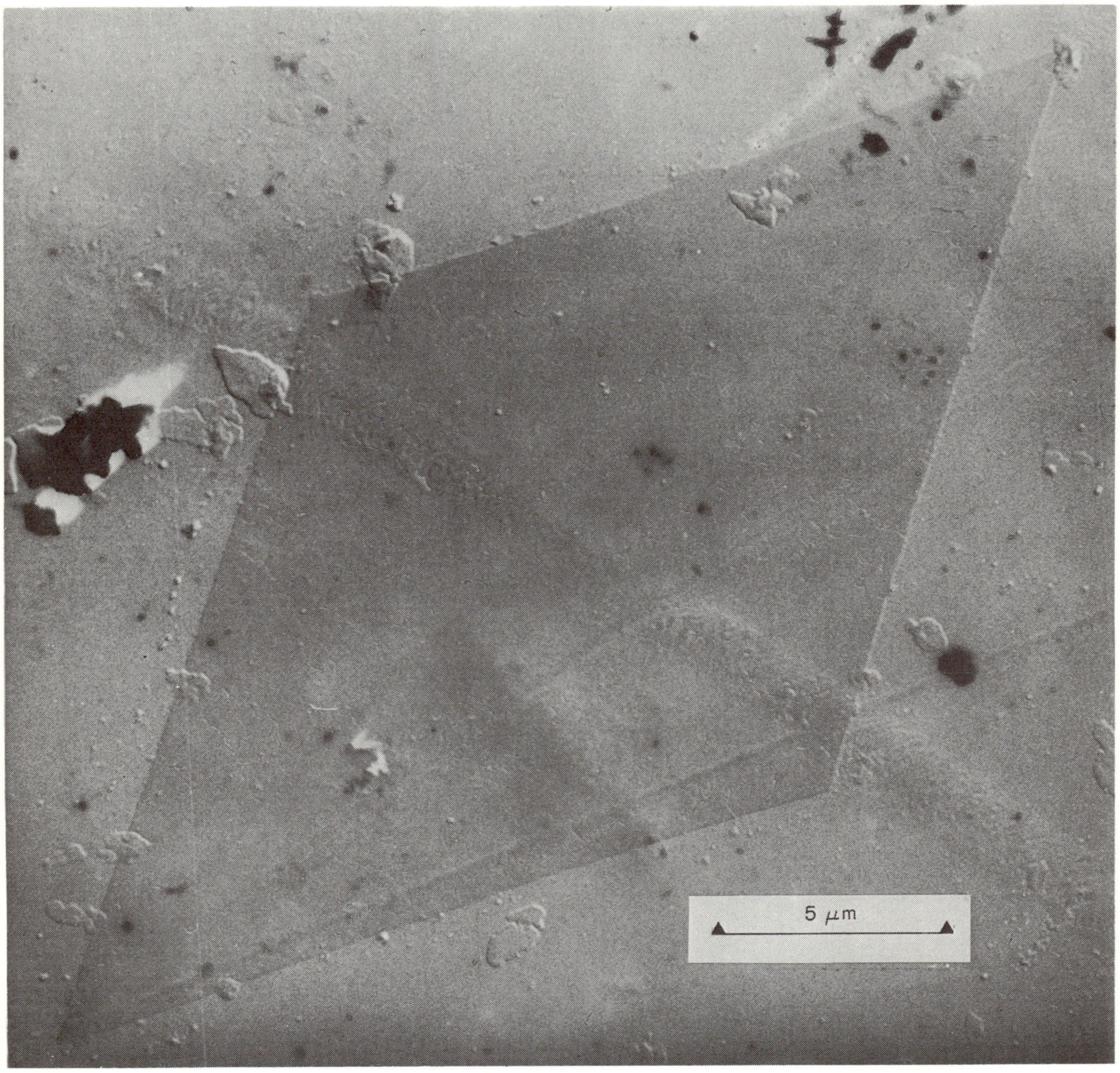

(a)

Figure 4.50 (a) A single crystal of deformed polyethylene. (b) SADP. (c) The positions of the spots in (b). (d) One indexed component of (b). (e) Two twin components of (b) [From Allan *et al.* (1973), courtesy of *The Philosophical Magazine*]

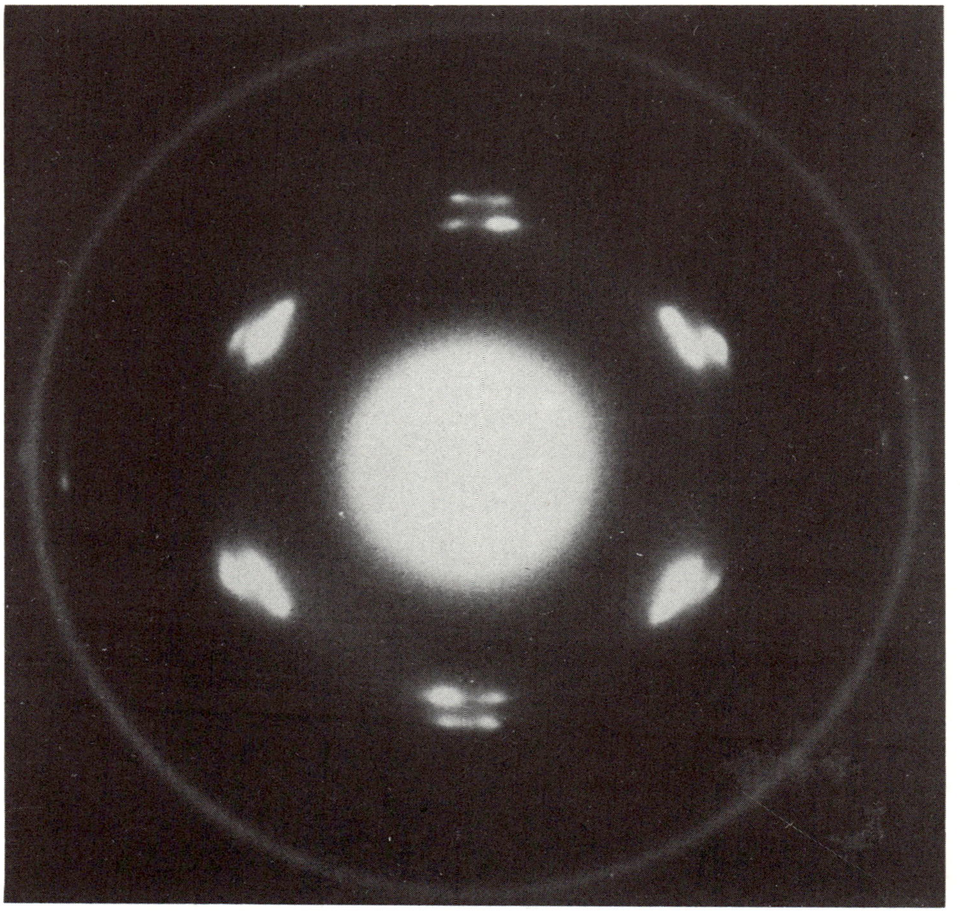

(b)

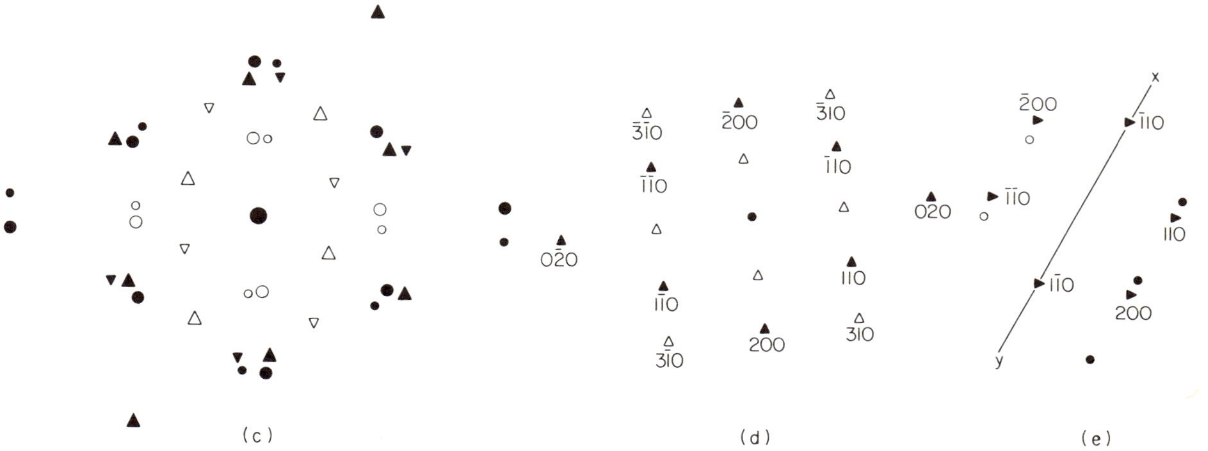

(c) (d) (e)

Figure 4.50 (*continued*)

distribution analysis of x-ray diffraction patterns (Bjornhang *et al.*, 1954) and transmission electron microscopy (Yeh and Geil, 1968; Yeh, 1972) can be interpreted in terms of regions of local molecular order. In this section an example is shown of the use of CDF techniques in the study of order in isotactic and atactic polystyrene.

Figure 4.51(a) shows a typical diffraction pattern obtained from a fresh amorphous isotactic polystyrene. Two diffuse rings are visible corresponding to reflecting plane spacings of 4.78 Å and 2.23 Å. Rings were also visible on the negative for spacings of 9 Å and 1.26 Å. Similar results are obtained for isotactic polystyrene. Figure 4.51(b),(c) shows CDF images obtained using a portion of the 4.78 Å ring of an atactic polystyrene. The bright regions have been interpreted (Yeh, 1972) as regions of local order. Since only a few ordered regions contribute to any segment of the ring included in the objective aperture, only a few ordered regions of the specimen are visible.

Direct measurements of randomly selected regions lead to a particle size of 15–40 Å for atactic specimen which is approximately 30% higher than that measured from line broadening (Yeh, 1972). These differences may be ascribed to the fact that phase contrast effects are important in the formation of these images, see section 3.18. Consequently, the image is influenced by defocussing, see section 1.5.5. Cowley (1973a,b) has pointed

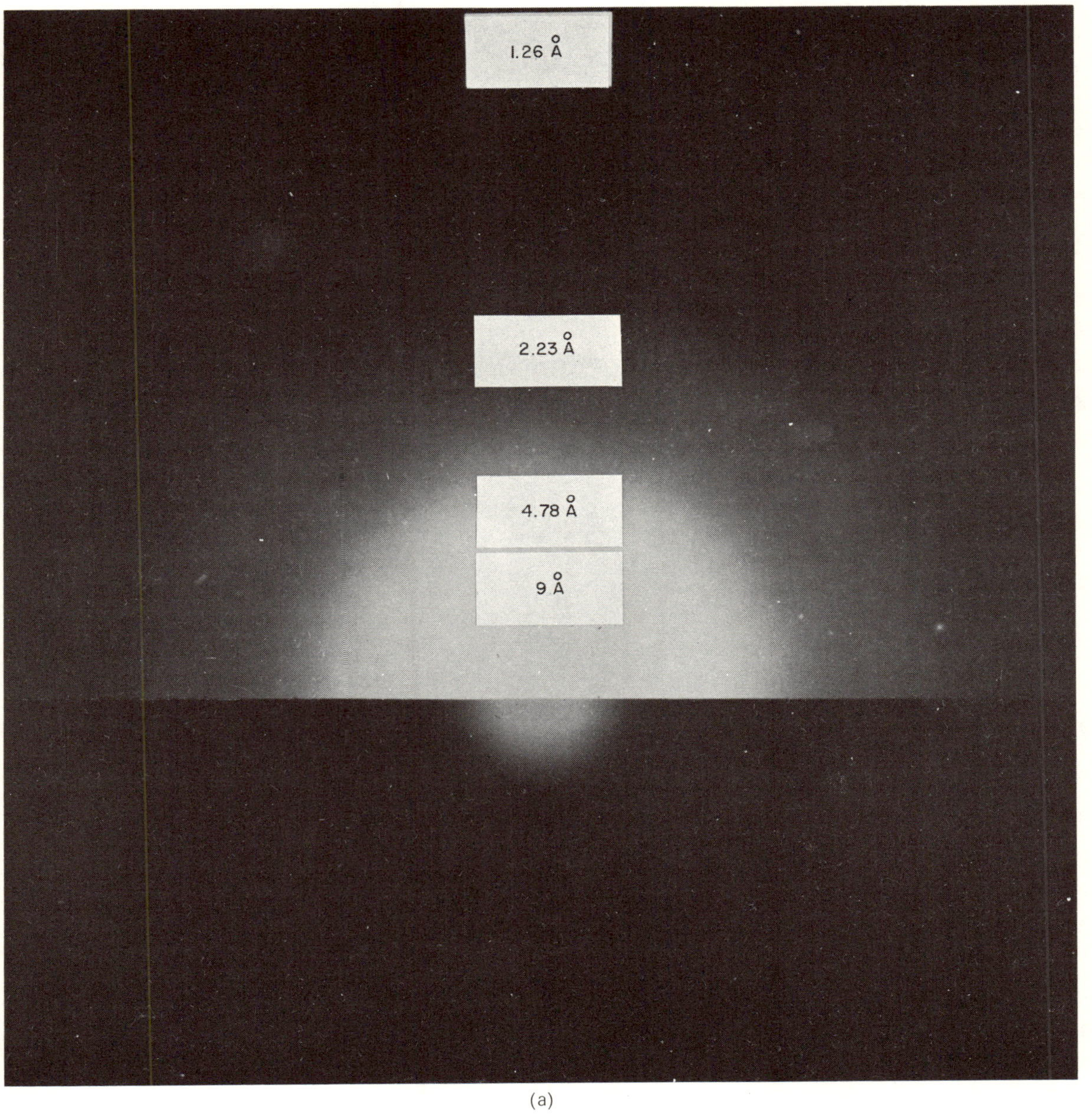

(a)

Figure 4.51 (a) A typical electron diffraction pattern of fresh amorphous, isotactic polystyrene. Atactic polystyrenes show similar 'amorphous' patterns. (b), (c) CDF electron micrographs of an unshadowed thin film of atactic polystyrene (mol wt 1 800 000), using a portion of the 4.78 Å ring in (a) [From Yeh (1972), courtesy of *Journal of Macromolecular Science*]

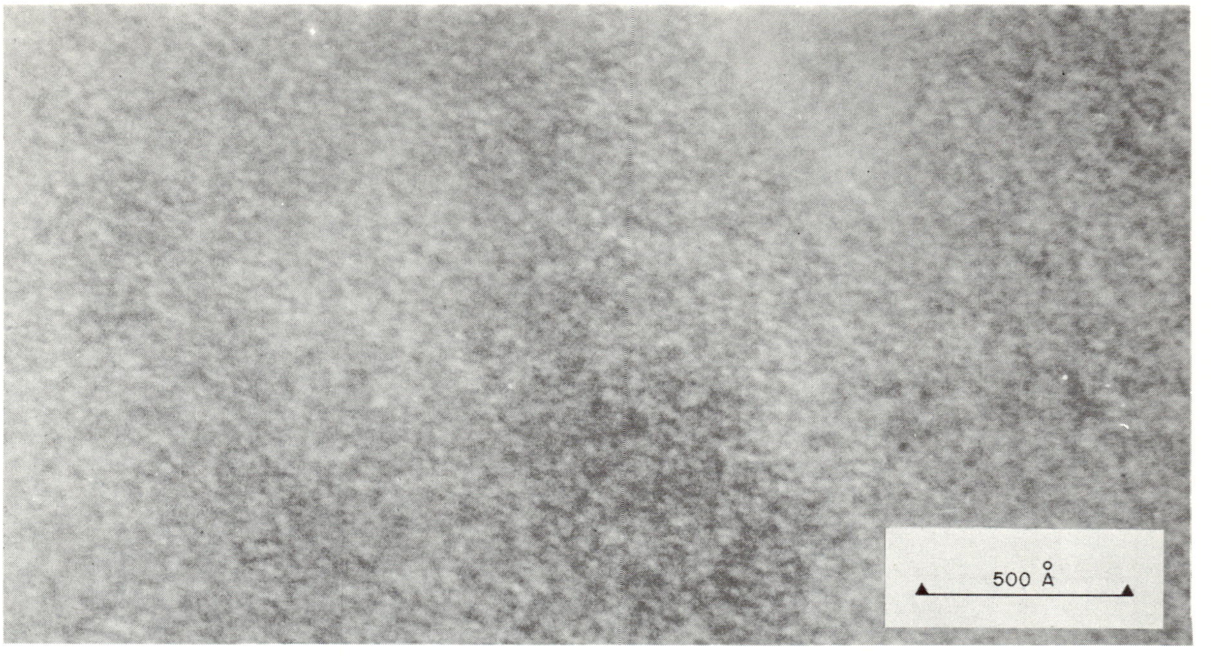

(b)

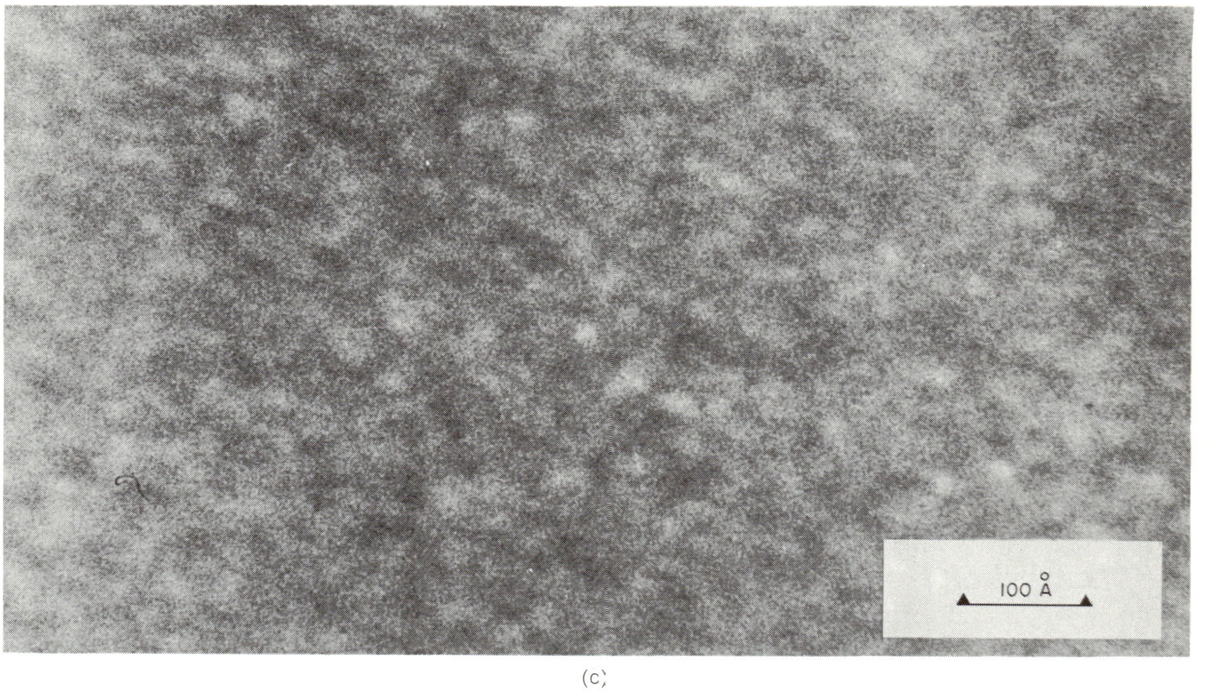

(c)

Figure 4.51 (*continued*)

out that interpretation of each white dot as an individual region of preferred atom correlation can be incorrect, particularly if the specimen is thicker than ~ 50 Å when overlap effects and the presence of background diffuse scattering can radically modify the image. Consequently, detailed interpretation of these images in terms of particle size and identity can be very difficult. For further details, see section 3.18.

However, combination of CDF images and SADP with other techniques such as x-ray analysis are likely to be important in the future in obtaining a better understanding of the structure of short-range ordered materials.

4.16 Semi-conductors

These materials have been the subject of a number

of studies using the electron microscope. Included in this section is an example of an investigation of ordering in amorphous semi-conductors.

4.16.1 *Amorphous Semi-conductor Materials (after M. L. Rudee)*

Currently there is considerable interest in the structure of amorphous semi-conductor materials. As for the case of polymers discussed in section 4.15.2, the conventional approach involves analysis of x-ray measurements which leads to radial distribution functions that may be discussed in terms of the structure of the semi-conductor. Transmission electron microscopy has not yet been applied to this problem in detail because of the difficulty of interpreting the image, see section 3.22. However, an example of the use of electron microscopy in this area is included here to show the type of investigation possible at present.

The structural models proposed for amorphous materials are of two types.

(1) Continuously random models where no regular short-range order is thought to exist.

(2) Domain models in which very small ordered domains are thought to be present.

Rudee (1972) has provided electron microscope evidence that the latter is a more accurate description of amorphous silicon and germanium. Figure 4.52(c) inset is a typical electron diffraction pattern obtained from an evaporated thin germanium film. The diffuse rings are characteristic of amorphous films and microdensitometer traces across the pattern shows no evidence of unevenness in the traces that might indicate the presence of crystallites. CDF images may be formed with specific segments of the diffuse rings, see figure 4.52(a),(b) which shows two such CDF images of the same area of a silicon film. As can be seen, different regions of the specimen are bright in the two cases. Rudee found that the results were independent of the method of making the film and the period of viewing of the electron microscope.

This type of image could arise from the interaction of the electron beam with a random structure, but this is thought to be unlikely for the two following reasons.

(1) The size of the spots in the image (20 Å and 15 Å diameter, Si and Ge, respectively) was independent of the objective aperture size.

(2) The objective aperture angle $\sim 0.3°$ was much less than the 10° band angle used in most random models. Thus it is unlikely that coherent diffraction could occur from regions 15 Å in diameter and still be passed through the objective aperture.

Additional indirect evidence such as whether or

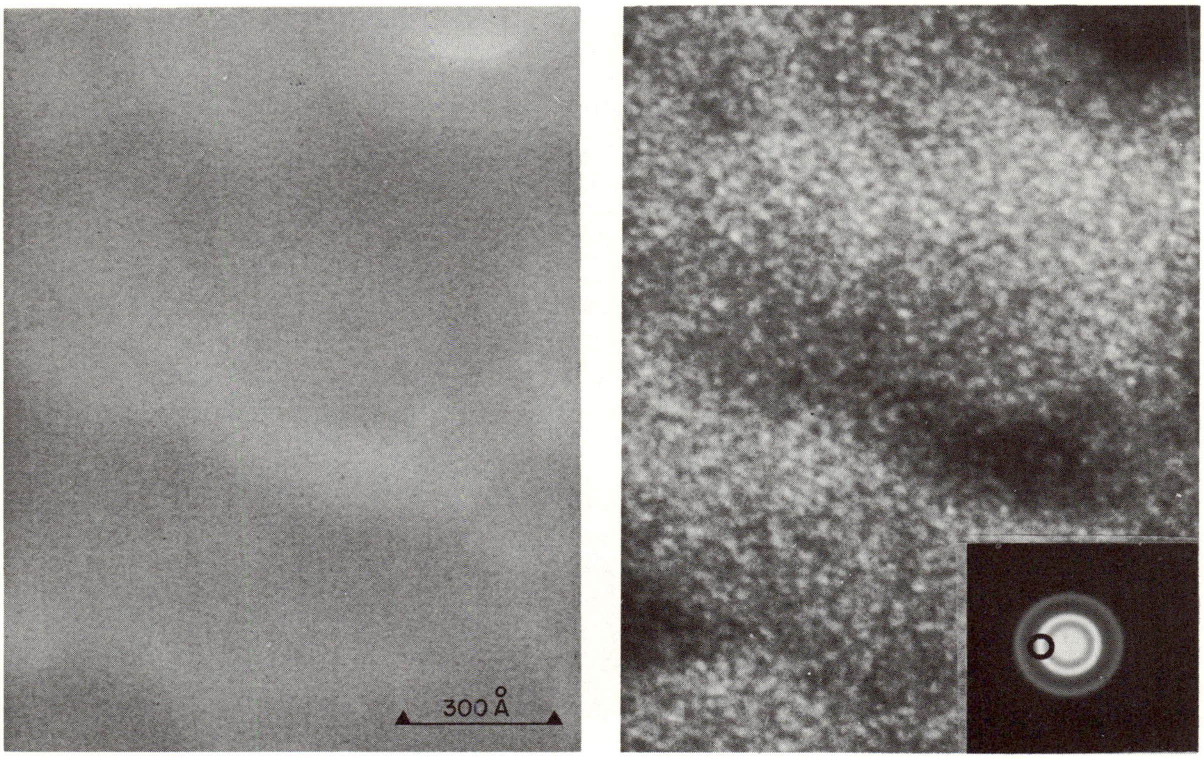

(a) (b)

Figure 4.52 (a), (b) CDF micrographs from the same area in an amorphous Si film. The circle indicates the region of the diffraction pattern passed by the objective aperture

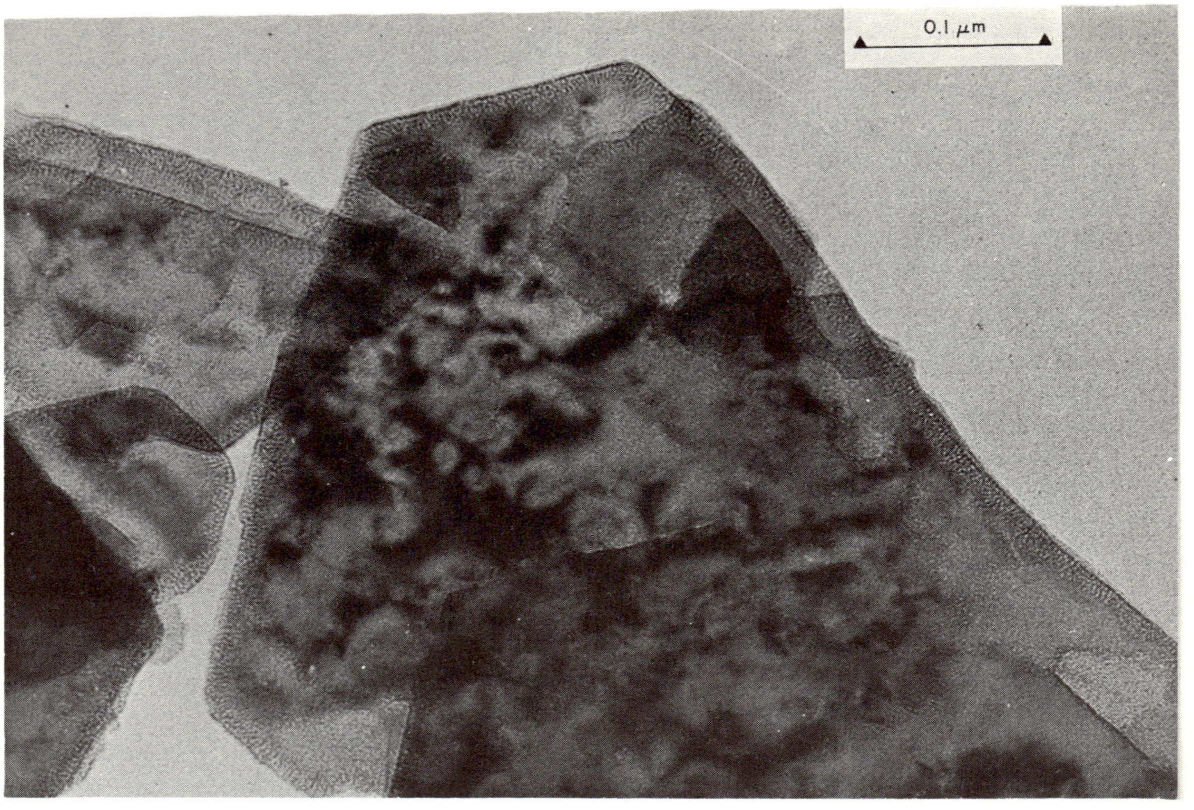

(a)

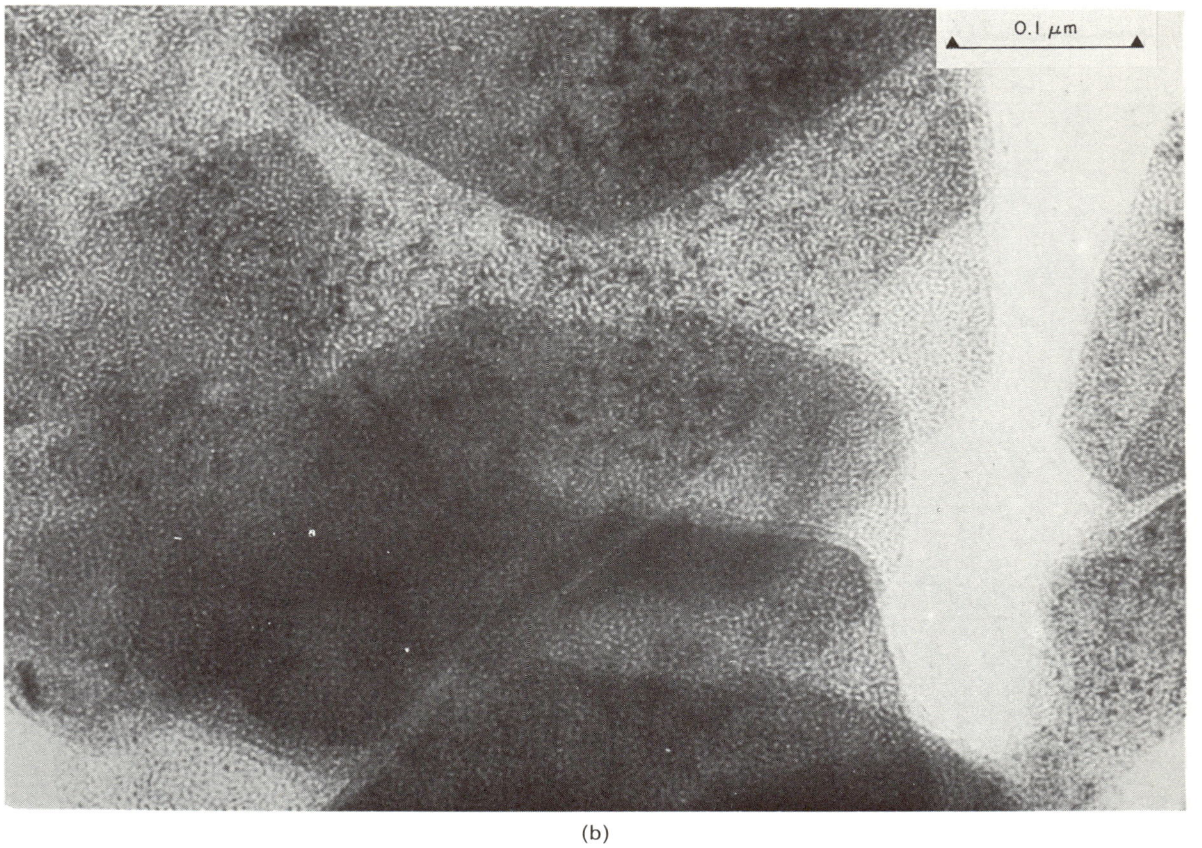

(b)

Figure 4.53 (a), (b) Direct observation of the dehydration of single-crystal flakes of $Co(OH)_2$, phase contrast (defocus) image. (c) A schematic diagram illustrating the mechanism of dehydration. (d) A phase contrast (defocus) image of slit-like pores in dehydrated $Al(OH)_3$ [From McConnell (1971), courtesy of *The Cambridge Review*]

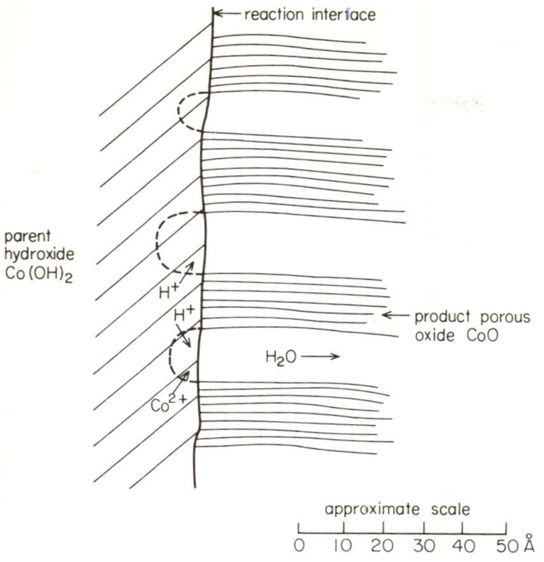

(c)

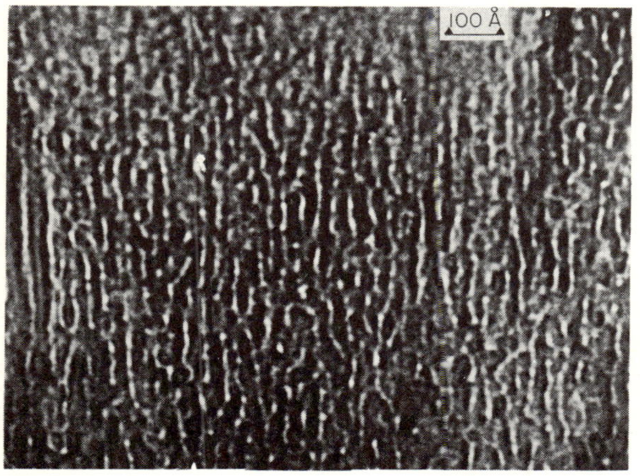

(d)

Figure 4.53 (*continued*)

not the specimen thickness could be accounted for by adding the bright dots in DF images implied that these images were real. Consequently, Rudee was able to conclude that this experiment supports the view that amorphous Ge is composed largely of ordered domains. However, it must be emphasised that definitive interpretation of the data requires the development of a more quantitative theory of image formation, taking into account phase contrast effects, as described in section 3.18, and by Cowley (1973a,b).

4.17 Minerals

Electron microscopy has been applied to a number of problems associated with minerals. In this section examples are included of investigations of porous catalysts and asbestos.

4.17.1 *Studies of Porous Catalysts* (*after J. D. C. McConnell*)

Porous solids for use as catalysts or catalyst supports are frequently produced by dehydration of hydroxides. These materials owe their importance in both applications to their large internal surface area on which adsorption and reaction can take place. The electron microscope enables the pore structure of these catalysts to be observed directly and the dehydration reactions to be followed *in situ*.

The direct observation of the dehydration reaction of $Co(OH)_2$ to form CoO is illustrated in figure 4.53(a),(b) using a single-crystal flake of the hydroxide, ~100 Å thick. The dehydration process begins at the edge of the specimen and progresses towards the centre. The CoO formed is polycrystalline with a high degree of preferred orientation. The reaction at the interface involves the advance of pores and the process may be readily explained within terms of figure 4.53(c). This illustrates counter migration of Co ions and protons in the region just ahead of the pore. Protons combine with hydroxyl ions to form water molecules which subsequently diffuse out through the pores. The form of the pores developed during dehydration depends on the material. Narrow, slit-like pores form during dehydration of $Al(OH)_3$ which are parallel to the original layer lattice, see figure 4.53(d).

In all of the images in figure 4.53(a),(b),(d) the phase contrast effects described in section 3.18 are responsible for contrast which is enhanced by a small degree of underfocus as described in section 1.5.5. The bright dots are interpreted as pores although it is recognised that this interpretation may not be unique because they could also arise from regions of material more dense than average as described in section 3.22 (Cowley, 1973a,b). However, the above interpretation is reinforced if the specimen is tilted into a strongly diffracting position when the pores are visible because of thickness contrast as described in section 3.12.

Pore size measurements using both phase and diffraction contrast images show that the minimum diameter is ~10 Å. However, this value is subject to errors arising from defocus and image transfer coefficients in the phase contrast image (see sections 3.18, 1.5.5) and chromatic aberration in the diffraction contrast image (see section 1.2). Consequently, detailed comparison between observed pore sizes and densities with those calculated from gas absorption analysis is necessary before detailed conclusions may be drawn about the surface area available for catalysis.

4.17.2 *High-resolution Electron Microscopy of Chrysotile Asbestos* (*after K. Yada*)

Chrysotile asbestos has a monoclinic crystal structure, $a = 5.3$ Å, $b = 9.1$ Å, $c = 7.32$ Å, $\alpha = \gamma = 90°$,

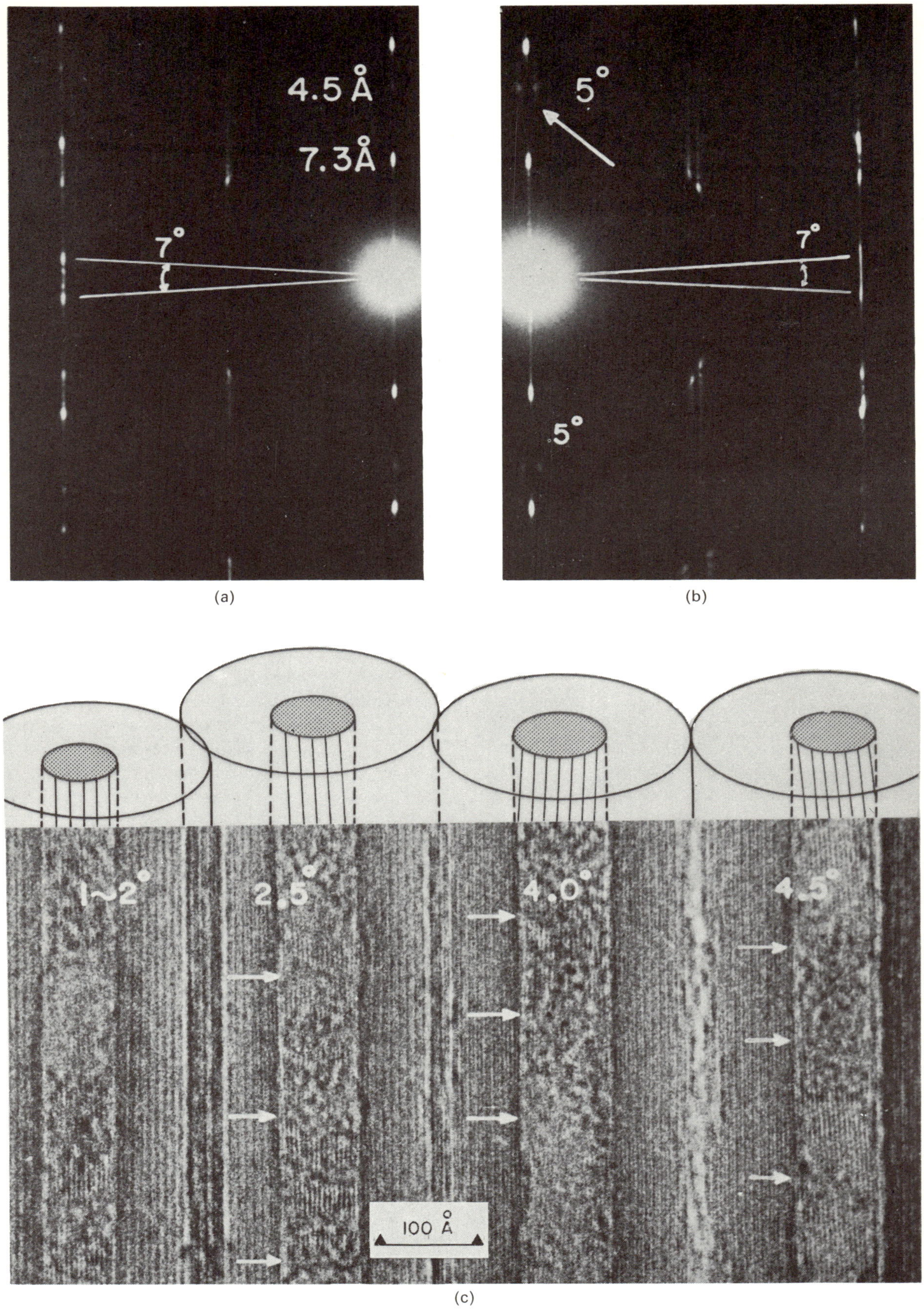

Figure 4.54 SADP from the fibrils showing (a) parallel fringes and (b) tilted fringes. (c) Four fibrils of chrysotile asbestos showing tilted 4.5 Å fringes in the middle of each fibril. The arrows indicate places where the contrast of the 4.5 Å fringes disappears by the Moiré effect [From Yada (1971), courtesy of *Acta Crystallographica*]

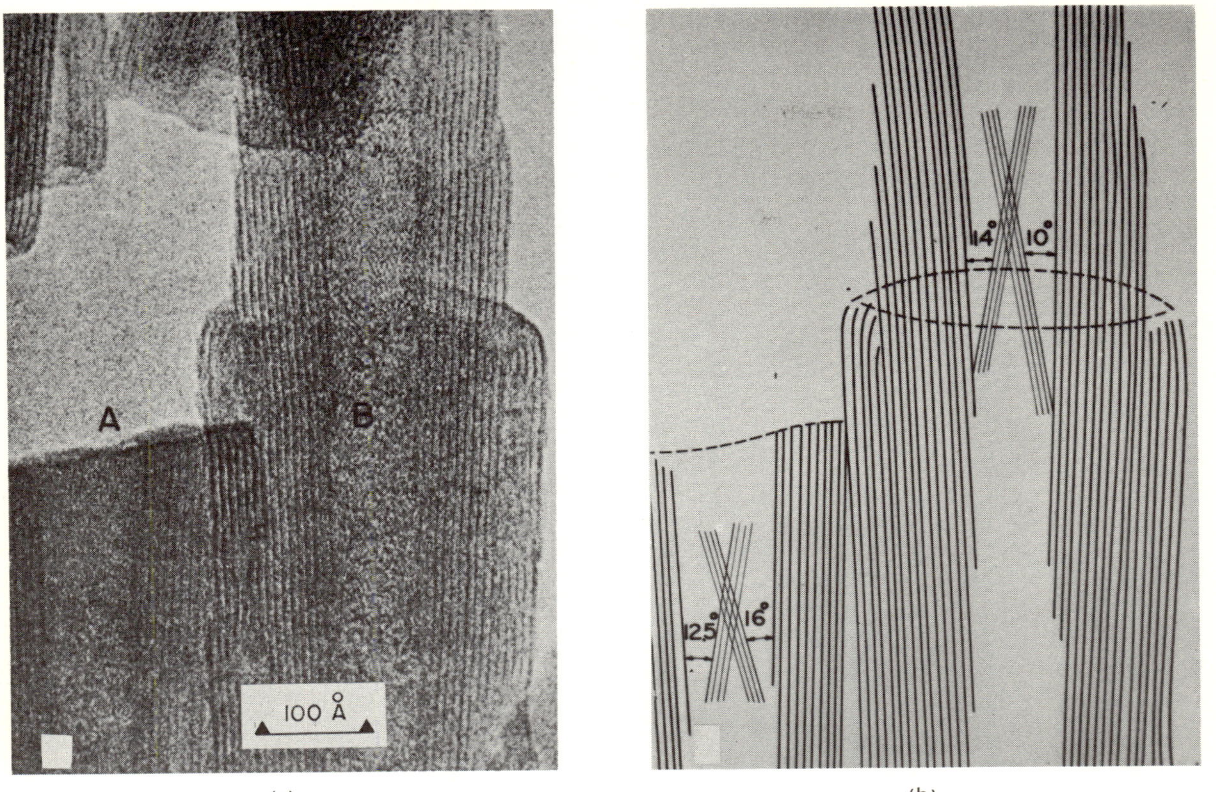

(a) (b)

Figure 4.55 Conical fibril in synthetic crysotile asbestos for which the 7.3 Å lattice fringes on opposite sides are not parallel to each other. The conically tapered angles are 3° (A) and 5.5° (B). (b) Is a diagram of the fringes [From Yada (1971), courtesy of *Acta Crystallographica*]

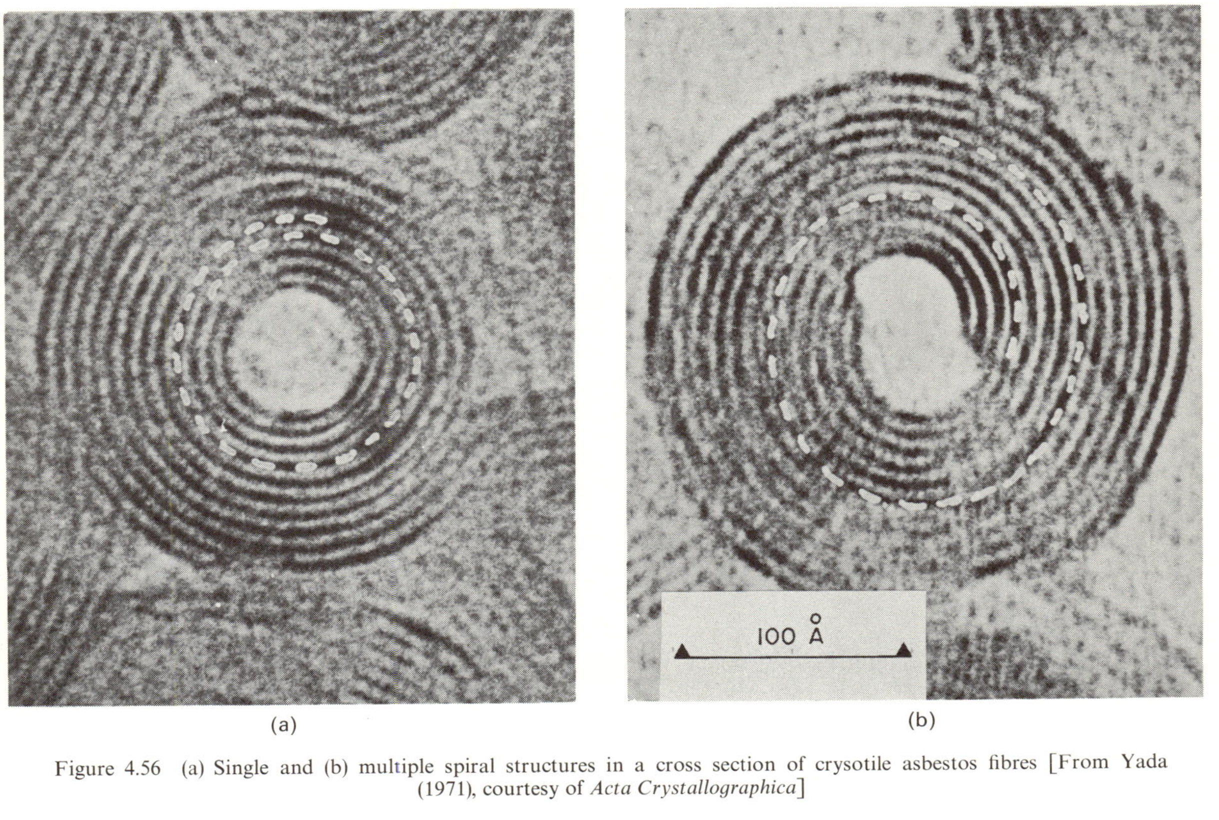

(a) (b)

Figure 4.56 (a) Single and (b) multiple spiral structures in a cross section of crysotile asbestos fibres [From Yada (1971), courtesy of *Acta Crystallographica*]

$\beta = 93°$, and consists of hollow fibrils of material. Several aspects of the morphology and structure of this material can be clarified using high-resolution lattice fringe electron microscopy. The details of the methods used to produce these images are described in section 1.5.5 and the physical reasons for their formation are summarised in section 3.19. In the examples described below, axial illumination and a pointed filament were used, together with effective anti-contamination and a low-beam current to minimise damage to the specimen.

The following examples indicate the type of information that may be obtained using these fringe images.

Figure 4.54(c) shows a typical high-resolution picture of four fibrils with fringes visible, spacing 4.5 Å, corresponding to (020). The diffraction pattern from one fibril with parallel fringes is shown in figure 4.54(a). It may be indexed as shown using the rules outlined in section 2.7.2.1. In figure 4.54(c) it is clear that some of the fringes in the centre of the fibril are tilted through the angles marked relative to its long axis. Furthermore, their (020) reflection, figure 4.54(b), is split into two spots (compare with figure 4.54(a) which is taken from fibrils with fringes parallel to the long axis). The $\{hko\}$ spots are also split and the angle of splitting for both reflections corresponds to twice that between the long axis of the fibril and the (020) fringes. This tilting and splitting is consistent with the interpretation of the crystal structure of the fibril as a helical role but a detailed interpretation is not available at present.

Conical fibrils are observed in synthetic chrysotile. Figure 4.55(a) shows 7.3 Å spacing fringes corresponding to (001) but these are not parallel on either side of the fibril. Figure 4.55(b) shows the geometrical interpretation in terms of interlocked cones reported by Bates and Comer (1959).

Samples cut and viewed parallel to the fibre axis provide useful information on the growth processes. Figure 4.56 shows that concentric single and multi-spiral growths can exist in the same specimen. These images provide direct evidence for the importance of the axial, radial and coupled dislocations suggested by Jagodzinski and Kunze (1954) to explain the growth mechanism of the hollow fibrils.

4: References

Allan, P., Crellin, E. B., and Bevis, M. (1973). *Phil. Mag.*, **27,** 127.

Andrews, K. W., Dyson, D. J., and Keown, S. R. (1971). *Interpretation of Electron Diffraction Patterns*, Hilger, London.

Ashbee, K. H. G., and Smallman, R. E. (1963). *Proc. Roy. Soc. (London), Ser. A*, **274,** 195.

Ashbee, K. H. G., Smallman, R. E., and Williamson, G. K. (1963). *Proc. Roy. Soc. (London), Ser. A*, **276,** 542.

Ashby, M. F. (1964). *Z. Metallk.*, **55,** 5.

Auld, J. H., and Cousland, S. McK. (1971). *Scripta Met.*, **5,** 765.

Bagaryatski, Y. A. (1950). *Dokl. Akad. Nauk SSSR*, **73,** 1161.

Baker, R. G., and Nutting, J. (1959). Precipitation processes in steels. *Iron and Steel Institute, Spec. Rept.*, No. 64.

Balcerzack, A. T., and Sass, S. L. (1972). *Met. Trans.*, **3,** 1601.

Bates, T. F., and Comer, J. J. (1959). *Clay and Clay Minerals* (*Proc. 6th Natl. Clay Conf.*), **6,** 237.

Beech, J., and Warrington, D. H. (1966). *J. Iron Steel Inst. (London)*, **204,** 460.

Bevis, M., and Crellin, E. B. (1971). *Polymer*, **12,** 666.

Billingham, J., Bell, P. S., and Lewis, M. H. (1972). *Phil. Mag.*, **25,** 661.

Bjornhang, A., Ellefsen, O., and Tonnesen, B. A. (1954). *J. Polymer Sci.*, **12,** 621.

Blackburn, M. J., and Williams, J. C. (1967). *Trans. AIME*, **239,** 287.

Boyd, J. D. (1969). *Quart. Trans. Am. Soc. Metals*, **62,** 977.

—— (1970). *The Science, Technology and Applications of Titanium* (eds. R. I. Jaffee and N. E. Promisel), Pergamon Press, Oxford, p. 545.

Bunn, C. W. (1939). *Trans. Faraday Soc.*, **35,** 482.

Clarebrough, L. M. (1971). *Aust. J. Phys.*, **24,** 79.

Cowley, J. M. (1973a). *Acta Cryst.*, **29,** 529.

—— (1973b). *Acta Cryst.*, **29,** 537.

Davenport, A. T. (1968). *J. Iron Steel Inst. (London)*, **206,** 499.

Davenport, A. T., Berry, F. G., and Honeycombe, R. W. K. (1968). *Met. Sci. J.*, **2,** 104.

Davenport, A. T., and Honeycombe, R. W. K. (1971). *Proc. Roy. Soc. (London), Ser. A*, **322,** 191.

Degischer, H. P. (1972). *Phil. Mag.*, **26,** 1137.

Dippenaar, R. J., and Honeycombe, R. W. K. (1973). *Proc. Roy. Soc. (London), Ser. A*, **333,** 455.

Dyson, D. J., and Andrews, K. W. (1969). *J. Iron Steel Inst. (London)*, **207,** 208.

Embury, J. D., and Nicholson, R. B. (1965). *Acta Met.*, **13,** 403.

Gjønnes, J., and Simensen, Chr. J. (1970). *Acta Met.*, **18,** 881.

Goldschmidt, H. J. (1967). *Interstitial Alloys*, Butterworths, London.

Grubb, D. T., and Groves, G. W. (1971). *Phil. Mag.*, **24,** 815.

Grubb, D. T., Keller, A., and Groves, G. (1972a). *J. Mater. Sci.*, **7,** 131.

—— (1972b). *J. Mater. Sci.*, **7,** 822.

Hannick, R. J., Murray, M. J., and Packer, M. E. (1971). *Phil. Mag.*, **24,** 1179.

Hay, I. L., and Keller, A. (1970). *J. Polymer Sci.*, **B4,** 481.

Head, A. K. (1967). *Aust. J. Phys.*, **20,** 557.

Hilliard, J. E. (1962). *Trans. AIME*, **224,** 906.

Howie, L. M., McGurn, J. F., and Gilbert, R. W. (1966). *Acta Met.*, **14,** 801.

Humble, P. (1968). *Aust. J. Phys.*, **21,** 325.

Jagodzinski, H., and Kunze, G. (1954). *Nues, Jahrb. Mineral. Mh.*, p. 137.

Jones, P. J., and Edington, J. W. (1972a). *Met. Trans.*, **3,** 2706.

—— (1972b). *Phil. Mag.*, **25,** 729.

—— (1973). *Phil. Mag.*, **27,** 393.

Kear, B. H., Leverant, G. R., and Oblak, J. M., 1969, *Quart. Trans. Am. Soc. Metals*, **62,** 639.

Keown, S. R., and Dyson, D. J. (1966). *J. Iron Steel Inst. (London)*, **204,** 832.

Kiho, H., Peterlin, A., and Geil, P. H. (1964). *J. Appl. Phys.*, **35,** 1599.

van Landuyt, J. (1966). *Phys. Stat. Sol.*, **16,** 585.

van Landuyt, J., Gevers, R., and Amelinckx, S. (1964). *Phys. Stat. Sol.*, **7,** 307.

—— (1965). *Phys. Stat. Sol.*, **9,** 135.

—— (1966a). *Phys. Stat. Sol.*, **13,** 467.

—— (1966b). *Phys. Stat. Sol.*, **18,** 167.

Lewis, M. H., Billingham, J., and Bell, P. S. (1972). *Electron Microscopy and Structure of Materials* (eds. G. Thomas, R. M. Fulrath and R. M. Fisher), University of California Press, Berkeley, California, p. 1084.

Loretto, M. H., Clarebrough, L. M., and Humble, P. (1966). *Phil. Mag.*, **13,** 953.

Loretto, M. H., and France, L. K. (1969). *Phil. Mag.*, **19,** 141.

McCabe, K. K., and Sass, S. L. (1971). *Phil. Mag.*, **23,** 957.

McConnell, J. D. C. (1971). *The Cambridge Review*, **8,** 12.

McIntyre, K. G., and Brown, L. M. (1966). *J. Phys. Radium*, **27,** C3–178.

Maher, D. M., and Eyre, B. L. (1971). *Phil. Mag.*, **23,** 409.

de Novion, C. H. (1966). *Compt. Rend.*, **263,** 775.

Oblak, J. M., and Kear, B. H. (1972). *Electron Microscopy and Structure of Materials* (eds. G. Thomas, R. M. Fulrath and R. M. Fisher), University of California Press, Berkeley, California, p. 566.

Partridge, P. G. (1967). *Met. Rev.*, **12,** No. 118, 169.

Pashley, D. W., and Presland, A. E. B. (1961). *Phil. Mag.*, **6,** 1003.

Pitsch, W. (1962). *Acta Met.*, **10,** 79 (errata, **10,** 906).

Rowcliffe, A. F., and Eyre, B. L. (1971). *J. Phys. F*, **1,** 771.

Rudee, M. L. (1972). *Electron Microscopy, and Strength of Materials* (eds. G. Thomas, R. M. Fulrath and R. M. Fisher), University of California Press, Berkeley, California, p. 1064.

Ryder, P. L., and Pitsch, W. (1968). *Phil. Mag.*, **18,** 807.

Sanderson, G., and Scully, J. C. (1967), *Trans. AIME*, **239,** 1883.

Sass, S. L. (1969), *Acta Met.*, **17,** 813.

—— (1971). *J. Appl. Cryst.*, **4,** 252.

Scheil, E. (1935). *Z. Metallk.*, **27,** 199.

Schmalzried, H., and Gerold, V. (1958), *Z. Metallk.*, **49,** 291.

Silcock, J. M., Davies, M. H., and Hardy, H. K. (1956). *Symp. on the Mechanism of Phase Transformations in Metals*, Institute of Metals, London, p. 93.

Silcock, J. M., and Denham, A. W. (1969). *The Mechanism of Phase Transformations in Crystalline Solids*, Institute of Metals, London, p. 59.

Silcock, J. M., and Tunstall, W. (1964). *Phil. Mag.*, **10,** 361.

Singhall, K., and Martin, J. W. (1968). *Trans. AIME*, **242,** 814.

Smallman, R. E. (1970). *Modern Physical Metallurgy*, Butterworths, London.

Speich, G. R. (1962). *Trans. AIME*, **224,** 850.

Tekin, E., and Kelly, P. M. (1965). *J. Iron Steel Inst. (London)*, **203,** 715.

Tufton, P. J. (1963). *Ph.D. Thesis*, University of Cambridge.

Tunstall, W. J. (1969). *Phil. Mag.*, **20,** 701.

Venables, J. D., Kahn, D., and Lye, R. G. (1968). *Phil. Mag.*, **18,** 177.

Westmacott, K. H., Smallman, R. E., and Dobson, P. S. (1968). *Met. Sci. J.*, **2,** 177.

Wilcox, B. A., and Clauer, A. H. (1969). *Met. Sci. J.*, **3,** 26.

Yada, K. (1971). *Acta Cryst.*, **A27,** 659.

Yakel, H. L. (1958). *Acta Cryst.*, **11,** 46.

Yeh, G. S. Y. (1972). *J. Macromol. Sci. Phys.*, **B6,** 451.

Yeh, G. S. Y., and Geil, P. H. (1968). *J. Macromol. Sci. Phys.*, **B2,** No. 1, 29.

Yoo, M. H., and Wei, C. T. (1967). *J. Appl. Phys.*, **38,** 4317.

APPENDIX 1. BASIC CRYSTALLOGRAPHY

A1.1 Introduction

In order to perform electron microscope studies it is necessary to understand the basic principles of crystallography set out below.

It is the regular arrangement of atoms in space which constitutes the distinguishing feature of the crystalline state. These regular arrays, called crystal structures, which are characterised in various different ways, give rise to the internal and external symmetry of the structure. The translational symmetry leads to the concept of a lattice (an array of atoms at points in space with identical surroundings). Arising from considerations of this type we arrive at fourteen distinct Bravais lattices, figure A1.1. These fall into the seven fundamental crystal classes listed in table A1.1. Most metals are either cubic, hexagonal or tetragonal in structure, see table A1.3. However, many non-metallic materials have more complex structures.

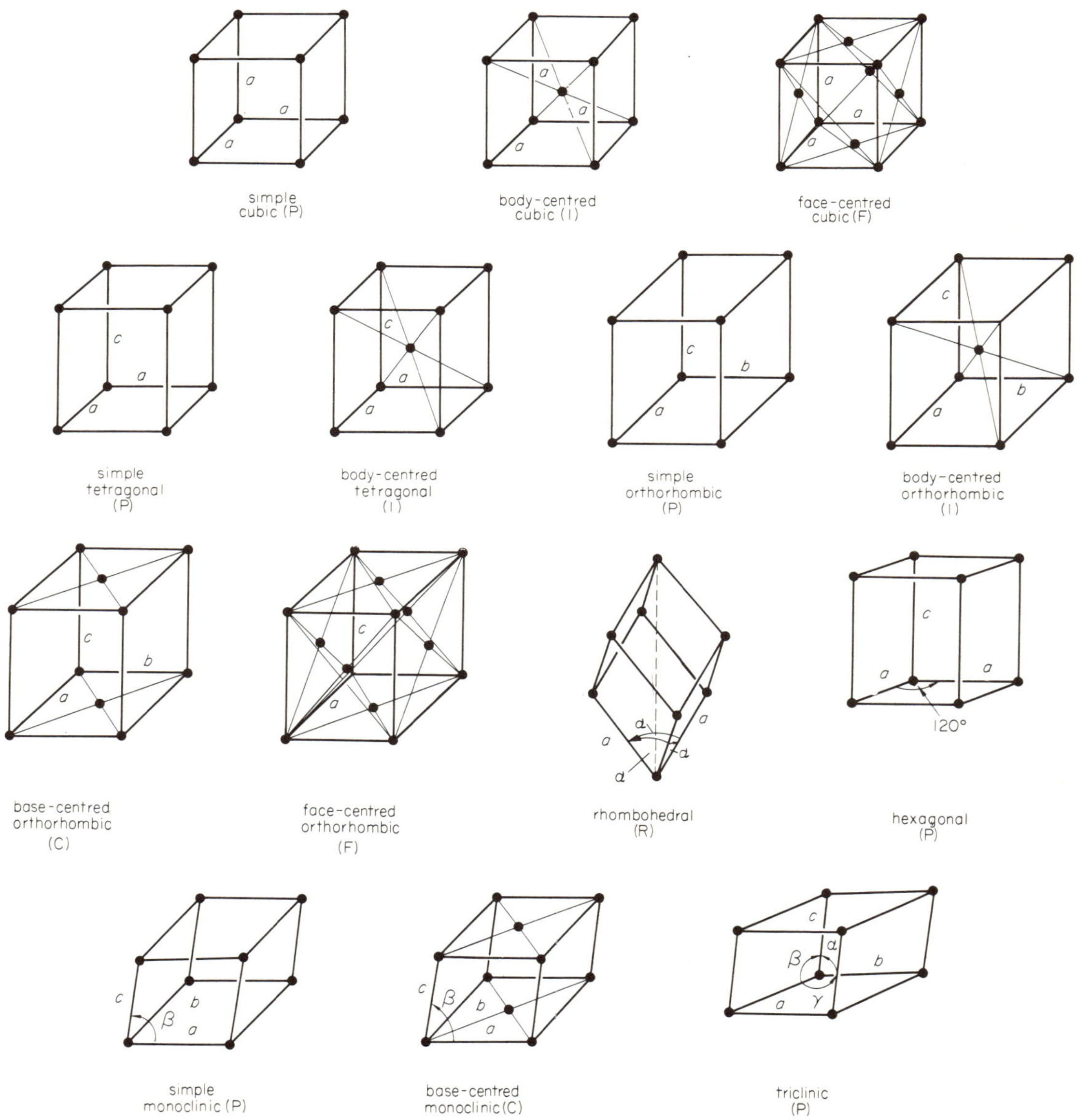

Figure A1.1 The fourteen Bravais lattices

Table A1.1 Crystallographic formulae for interplanar spacings,

Crystal system		Interplanar spacing of the (*hkl*) plane
cubic	$a = b = c$ $\alpha = \beta = \gamma = 90°$	$\frac{1}{d^2} = \frac{1}{a^2}(h^2 + k^2 + l^2)$
tetragonal	$a = b \neq c$ $\alpha = \beta = \gamma = 90°$	$\frac{1}{d^2} = \frac{1}{a^2}(h^2 + k^2) + \frac{1}{c^2}l^2$
orthorhombic	$a \neq b \neq c$ $\alpha = \beta = \gamma = 90°$	$\frac{1}{d^2} = \frac{1}{a^2}h^2 + \frac{1}{b^2}k^2 + \frac{1}{c^2}l^2$
hexagonal	$a = b \neq c$ $\alpha = \beta = 90°; \gamma = 120°$	$\frac{1}{d^2} = \frac{4}{3a^2}(h^2 + hk + k^2) + \frac{1}{c^2}l^2$
rhombohedral	$a = b = c$ $\alpha = \beta = \gamma < 120° \neq 90°$	$\frac{1}{d^2} = \frac{1}{a^2}\frac{(1 + \cos\alpha)\{(h^2 + k^2 + l^2) - (1 - \tan^2\frac{1}{2}\alpha)(hk + kl + lh)\}}{1 + \cos\alpha - 2\cos^2\alpha}$
monoclinic	$a \neq b \neq c$ $\alpha = \gamma = 90° \neq \beta$	$\frac{1}{d^2} = \frac{1}{a^2}\frac{h^2}{\sin^2\beta} + \frac{1}{b^2}k^2 + \frac{1}{c^2}\frac{l^2}{\sin^2\beta} - \frac{2hl\cos\beta}{ac\sin^2\beta}$
triclinic	$a \neq b \neq c$ $\alpha \neq \beta \neq \gamma$	$\frac{1}{d^2} = \frac{1}{V^2}(s_{11}h^2 + s_{22}k^2 + s_{33}l^2 + 2s_{12}hk + 2s_{23}kl + 2s_{33}lh)$ where $V^2 = a^2b^2c^2(1 - \cos^2\alpha - \cos^2\beta - \cos^2\gamma + 2\cos\alpha\cos\beta\cos\gamma)$ and $s_{11} = b^2c^2\sin^2\alpha$ $s_{22} = a^2c^2\sin^2\beta$ $s_{33} = a^2b^2\sin^2\gamma$ $s_{12} = abc^2(\cos\alpha\cos\beta - \cos\gamma)$ $s_{23} = a^2bc(\cos\beta\cos\gamma - \cos\alpha)$ $s_{31} = ab^2c(\cos\gamma\cos\alpha - \cos\beta)$

A1.2 Indexing Planes

Planes in any one of the fourteen Bravais lattices can be indexed in the same way. Axes are chosen defining a unit cell x, y and z at angles α, β and γ with the unit translation distances a, b and c (figure A1.2(a)).

A plane is defined in terms of its intercepts on these axes. For example, in figure A1.2(b) the plane cuts the axes at a/h, b/k and c/l where h, k and l are the Miller indices of the plane. Thus, in figure A1.2(c), the plane cuts the axes at unit translation distances in the x and y axes and two translation distances in the z axis. The index of the plane is then OX/OX, OY/OY and OZ/OW, that is $11\frac{1}{2}$ which is rationalised to 221, the *hkl* indices for the plane, that is the intercepts in the plane are a/h, b/k and c/l. A negative intercept results in a negative component of the Miller index written as $\bar{h}$.

Formulae for the angles between different (*hkl*) planes in the fourteen Bravais lattices are given in table A1.1.

A1.3 Indexing Lattice Directions

For any Bravais lattice, such as that shown in figure A1.3, the direction OA has the indices [121], that is the path from O to A involves moving 1 unit translation parallel to x, 2 parallel to y and 1 parallel to z. The other directions $[01\bar{1}]$ and $[\bar{1}\bar{1}0]$ are self-evident. The symbol $\bar{1}$ means one translation in the negative direction. The general symbol angles, and angles between directions for the seven crystal systems [After Andrews *et al.* (1971)]

Angle ϕ between $(h_1k_1l_1)$ and $(h_2k_2l_2)$	Angle ρ between directions $[u_1v_1w_1]$ and $[u_2v_2w_2]$
$\cos\phi = \dfrac{h_1h_2 + k_1k_2 + l_1l_2}{\{(h_1^2 + k_1^2 + l_1^2)(h_2^2 + k_2^2 + l_2^2)\}^{1/2}}$	$\cos\rho = \dfrac{u_1u_2 + v_1v_2 + w_1w_2}{\{(u_1^2 + v_1^2 + w_1^2)(u_2^2 + v_2^2 + w_2^2)\}^{1/2}}$
$\cos\phi = \dfrac{\frac{1}{a^2}(h_1h_2 + k_1k_2) + \frac{1}{c^2}l_1l_2}{\left[\left\{\frac{1}{a^2}(h_1^2 + k_1^2) + \frac{1}{c^2}l_1^2\right\}\left\{\frac{1}{a^2}(h_2^2 + k_2^2) + \frac{1}{c^2}l_2^2\right\}\right]^{1/2}}$	$\cos\rho = \dfrac{a^2(u_1u_2 + v_1v_2) + c^2w_1w_2}{[\{a^2(u_1^2 + v_1^2) + c^2w_1^2\}\{a^2(u_2^2 + v_2^2) + c^2w_2^2\}]^{1/2}}$
$\cos\phi = \dfrac{\frac{1}{a^2}h_1h_2 + \frac{1}{b^2}k_1k_2 + \frac{1}{c^2}l_1l_2}{\left\{\left(\frac{1}{a^2}h_1^2 + \frac{1}{b^2}k_1^2 + \frac{1}{c^2}l_1^2\right)\left(\frac{1}{a^2}h_2^2 + \frac{1}{b^2}k_2^2 + \frac{1}{c^2}l_2^2\right)\right\}^{1/2}}$	$\cos\rho = \dfrac{a^2u_1u_2 + b^2v_1v_2 + c^2w_1w_2}{\{(a^2u_1^2 + b^2v_1^2 + c^2w_1^2)(a^2u_2^2 + b^2v_2^2 + c^2w_2^2)\}^{1/2}}$
$\cos\phi = \dfrac{h_1h_2 + k_1k_2 + \frac{1}{2}(h_1k_2 + k_1h_2) + \frac{3}{4}\frac{a^2}{c^2}l_1l_2}{\left\{\left(h_1^2 + k_1^2 + h_1k_1 + \frac{3}{4}\frac{a^2}{c^2}l_1^2\right)\left(h_2^2 + k_2^2 + h_2k_2 + \frac{3}{4}\frac{a^2}{c^2}l_2^2\right)\right\}^{1/2}}$	$\cos\rho = \dfrac{u_1u_2 + v_1v_2 - \frac{1}{2}(u_1v_2 + v_1u_2) + \frac{c^2}{a^2}w_1w_2}{\left\{\left(u_1^2 + v_1^2 - u_1v_1 + \frac{c^2}{a^2}w_1^2\right) \times \left(u_2^2 + v_2^2 - u_2v_2 + \frac{c^2}{a^2}w_2^2\right)\right\}^{1/2}}$
convert to corresponding hexagonal indices (see appendix 2) and use the above two formulae	
$\cos\phi = \dfrac{\frac{1}{a^2}h_1h_2 + \frac{1}{b^2}k_1k_2\sin^2\beta + \frac{1}{c^2}l_1l_2 - \frac{1}{ac}(l_1h_2 + l_2h_1)\cos\beta}{\left\{\left(\frac{1}{a^2}h_1^2 + \frac{1}{b^2}k_1^2\sin^2\beta + \frac{1}{c^2}l_1^2 - \frac{2h_1l_1}{ac}\cos\beta\right) \times \left(\frac{1}{a^2}h_2^2 + \frac{1}{b^2}k_2^2\sin^2\beta + \frac{1}{c^2}l_2^2 - \frac{2h_2l_2}{ac}\cos\beta\right)\right\}^{1/2}}$	$\cos\rho = \dfrac{a^2u_1u_2 + b^2v_1v_2 + c^2w_1w_2 + ac(w_1u_2 + u_1w_2)\cos\beta}{\{(a_2u_1^2 + b^2v_1^2 + c^2w_1^2 + 2acu_1w_1\cos\beta) \times (a^2u_2^2 + b^2v_2^2 + c^2w_2^2 + 2acu_2w_2\cos\beta)\}^{1/2}}$
$\cos\phi = \dfrac{F}{A_{h_1k_1l_1}A_{h_2k_2l_2}}$ where $F = h_1h_2b^2c^2\sin^2\alpha + k_1k_2a^2c^2\sin^2\beta + l_1l_2a^2b^2\sin^2\gamma + abc^2(\cos\alpha\cos\beta - \cos\gamma)(k_1h_2 + h_1k_2) + ab^2c(\cos\gamma\cos\alpha - \cos\beta)(h_1l_2 + l_1h_2) + a^2bc(\cos\beta\cos\gamma - \cos\alpha)(k_1l_2 + l_1k_2)$ and $A_{hkl} = \{h^2b^2c^2\sin^2\alpha + k^2a^2c^2\sin^2\beta + l^2a^2b^2\sin^2\gamma + 2hkabc^2(\cos\alpha\cos\beta - \cos\gamma) + 2hlab^2c(\cos\gamma\cos\alpha - \cos\beta) + 2kla^2bc(\cos\beta\cos\gamma - \cos\alpha)\}^{1/2}$	$\cos\rho = \dfrac{L}{l_{u_1v_1w_1}l_{u_2v_2w_2}}$ where $L = a^2u_1u_2 + b^2v_1v_2 + c^2w_1w_2 + bc(v_1w_2 + w_1v_2)\cos\alpha + ac(w_1u_2 + u_1w_2)\cos\beta + ab(u_1v_2 + v_1u_2)\cos\gamma$ and $l_{uvw} = (a^2u^2 + b^2v^2 + c^2w^2 + 2bcvw\cos\alpha + 2cawu\cos\beta + 2abuv\cos\gamma)^{1/2}$

for a direction is $[uvw]$ where $\boldsymbol{r}$ = OA, figure A1.3, $= u\boldsymbol{a} + v\boldsymbol{b} + w\boldsymbol{c}$.

Consequently, $\boldsymbol{r}$ is a vector with components u, v and w along the axes.

Formulae for the angles between different $[uvw]$ directions in the fourteen Bravais lattices are given in table A1.1.

A1.4 Plane Normals

In the cubic crystal system only the direction normal to the plane (hkl) has indices $[hkl]$, for example [111] is the (111) plane normal. In all other crystal systems this is not true and table A1.2 gives the formulae for determining the indices of the directions $[uvw]$ normal to the plane (hkl) and vice versa.

A1.5 Zones and the Zone Law

Any two lattice planes intersect in a line which can be defined by the directional indices $[uvw]$. This is the axis for a prism of planes with this common direction. The planes are known as the zone of planes and the long axis is the zone axis, z, given the symbol $[UVW]$. The zone indices for any pair of planes, that is $(h'k'l')$ and (hkl) can be obtained in the following way:

$$\begin{array}{c|ccccc|c} h & k & & l & & h & k & l \\ & & \times & & \times & & \times & \\ h' & k' & & l' & & h' & k' & l' \end{array}$$

that is $U = kl' - k'l$, $V = lh' - l'h$, $W = hk' - h'k$. This is the cross product between plane indices or the directions of the plane normal.

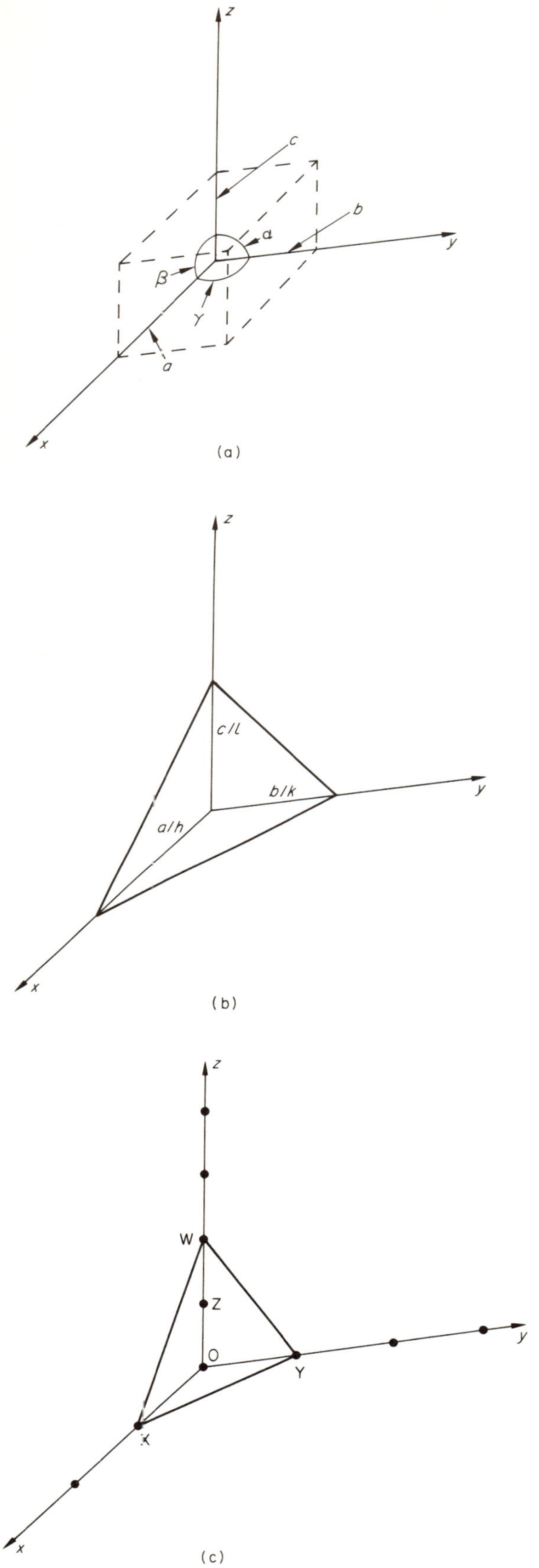

Figure A1.2 (a) The axes x, y, z defining the unit cell (dashed lines). (b) A general plane intersecting the axes. (c) A (221) plane

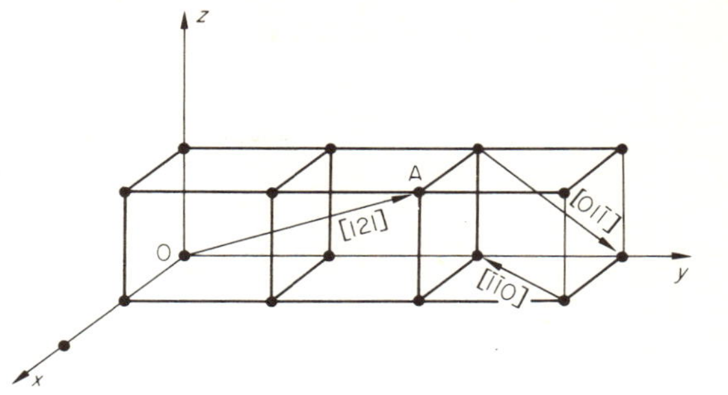

Figure A1.3 Crystal directions, OA is [121]

If it is necessary to find out if a plane (hkl) lies in a zone $[UVW]$ then the condition is

$$hU + kV + lW = 0$$

and this is called the Weiss zone law. This is essentially the condition that the normal to the plane (hkl) is perpendicular to $[UVW]$ direction.

Notation is as follows:

(111) means single set of parallel planes
{111} means equivalent planes of the type, that is (111), $(1\bar{1}1)$, etc.
[111] means a single zone axis or direction
⟨111⟩ means directions of equivalent type

A1.6 Stereographic Projection

Although other methods of projection of the three-dimensional crystal into two dimensions exist, the stereographic projection is the most common way of describing crystals. This is because the projection preserves angular truth. The advantage of working with the stereographic projection lies in the ease and rapidity of performing those crystallographic analyses necessary in the electron microscope.

Imagine a crystal at the centre of a sphere, see figure A1.4, with plane normals drawn from the centre of the sphere to its surface. In figure A1.5, one of these plane normals P is shown projected into the equatorial plane about the south pole of the sphere. Normally only those planes above the equator are projected. For planes underneath, the north pole is used as the projection point, indicated by open circles in the projection. The stereographic projection of the cubic crystal in figure A1.4 with [001] parallel to the south–north direction SN and [010] parallel to OD, is shown in figure A1.6, each point being indexed as the normal to a particular plane.

Standard projections for cubic crystal structures are shown in figure A1.12 with different planes in the centre of the stereogram.

It is necessary to be able to measure the angles between planes using the stereographic projection. The normals to any two crystal planes, for example

Table A1.2 Formulae defining the indices of the direction [*uvw*] perpendicular to plane (*hkl*) for the seven crystal systems [After Andrews *et al.* (1971)]

Crystal system	Equations for finding [*uvw*] given (*hkl*)	Equations for finding (*hkl*) given [*uvw*]
cubic	$\frac{u}{h} = \frac{v}{k} = \frac{w}{l}$	$\frac{h}{u} = \frac{k}{v} = \frac{l}{w}$
tetragonal	$\frac{u}{h} = \frac{v}{k} = \frac{w}{l}\frac{(c)^2}{(a)^2}$	$\frac{h}{u} = \frac{k}{v} = \frac{l}{w(c/a)^2}$
orthorhombic	$\frac{u}{h}a^2 = \frac{v}{k}b^2 = \frac{w}{l}c^2$	$\frac{h}{ua^2} = \frac{k}{vb^2} = \frac{l}{wc^2}$
hexagonal	$\frac{u}{2k + h} = \frac{v}{h + 2k} = \frac{2w}{3l}\frac{(c)^2}{(a)^2}$	$\frac{h}{2u - v} = \frac{k}{2v - u} = \frac{l}{2w(c/a)^2}$
rhombohedral	$\frac{u}{h \sin^2 \alpha + (k + l)(\cos^2 \alpha - \cos \alpha)} = \frac{v}{k \sin^2 \alpha + (l + h)(\cos^2 \alpha - \cos \alpha)}$ $= \frac{w}{l \sin^2 \alpha + (h + k)(\cos^2 \alpha - \cos \alpha)}$	$\frac{h}{u + (v + w) \cos \alpha} = \frac{k}{v + (w + u) \cos \alpha} = \frac{l}{w + (u + v) \cos \alpha}$
monoclinic	$\frac{u}{hb^2c^2 - lab^2c \cos \beta} = \frac{v}{kc^2a^2 \sin^2 \beta} = \frac{w}{la^2b^2 - hab^2c \cos \beta}$	$\frac{h}{ua^2 + wca \cos \beta} = \frac{k}{vb^2} = \frac{l}{uca \cos \beta + wc^2}$
triclinic	$\frac{u}{hs_{11} + ks_{12} + ls_{13}} = \frac{v}{hs_{12} + ks_{22} + ls_{23}} = \frac{w}{hs_{13} + ks_{23} + ls_{33}}$ ($s_{11} = b^2c^2 \sin^2 \alpha$, etc; $s_{12} = s_{21} = abc^2 (\cos \alpha \cos \beta - \cos \gamma)$, etc.)	$\frac{h}{ua^2 + vab \cos \gamma + wca \cos \beta} = \frac{k}{uab \cos \gamma + vb^2 + wbc \cos \alpha}$ $= \frac{l}{uca \cos \beta + vbc \cos \alpha + wc^2}$

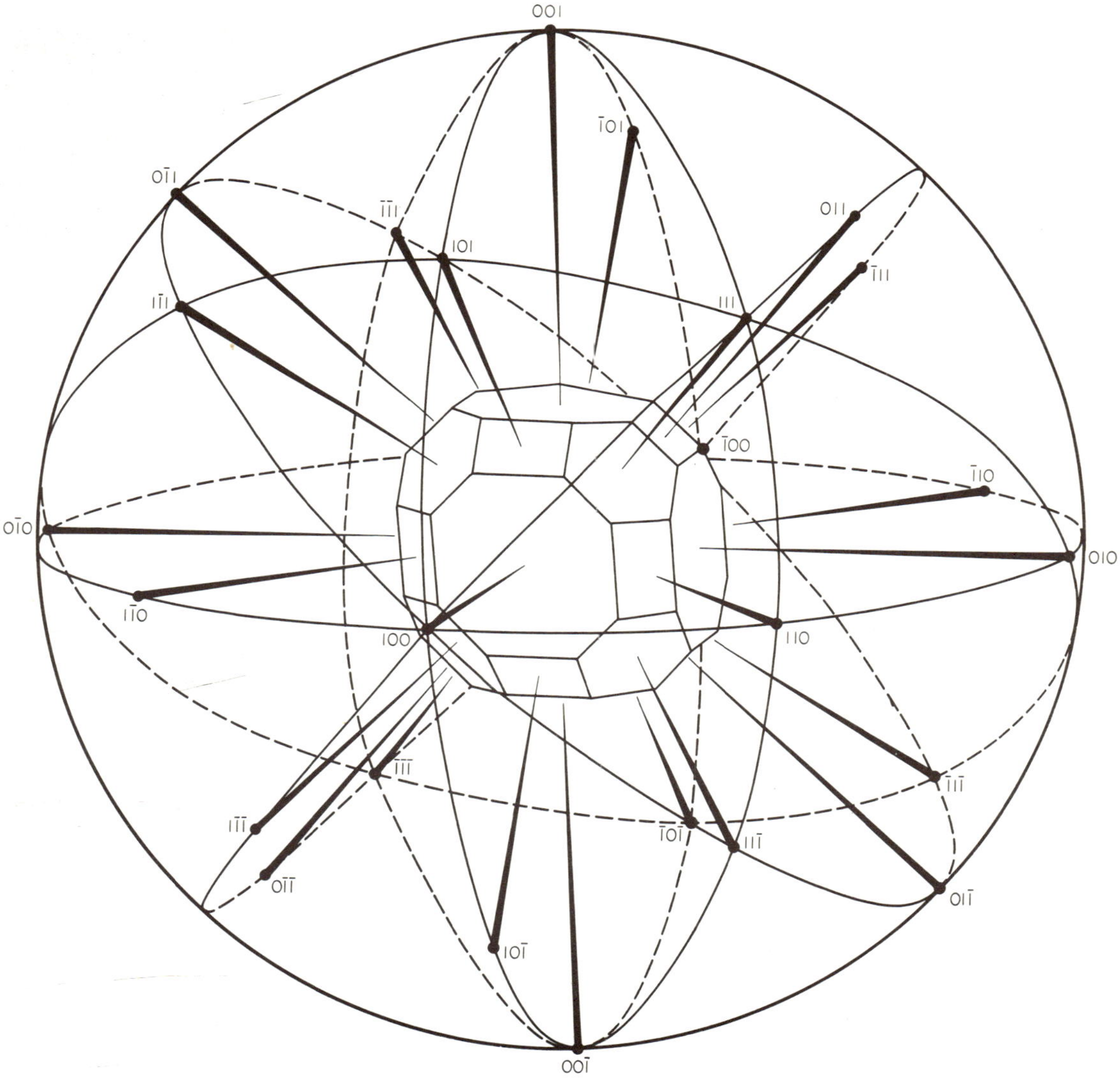

Figure A1.4 A crystal with cubic crystal structure situated at the centre of a sphere

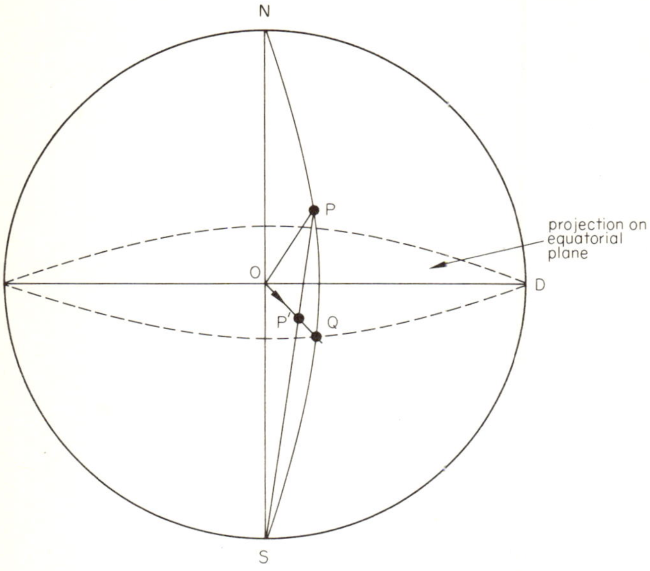

Figure A1.5 The projection of a plane normal OP into the equatorial plane about the south pole S

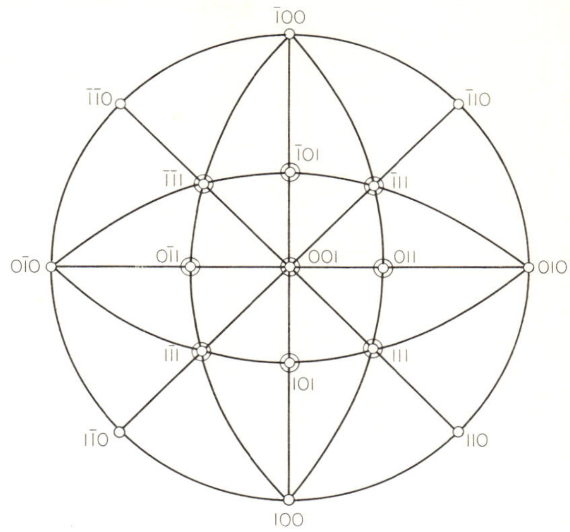

Figure A1.6 The stereographic projection for a cube crystal with [001] parallel to the south north direction in the sphere of figure A1.4. [010] is parallel to east–west

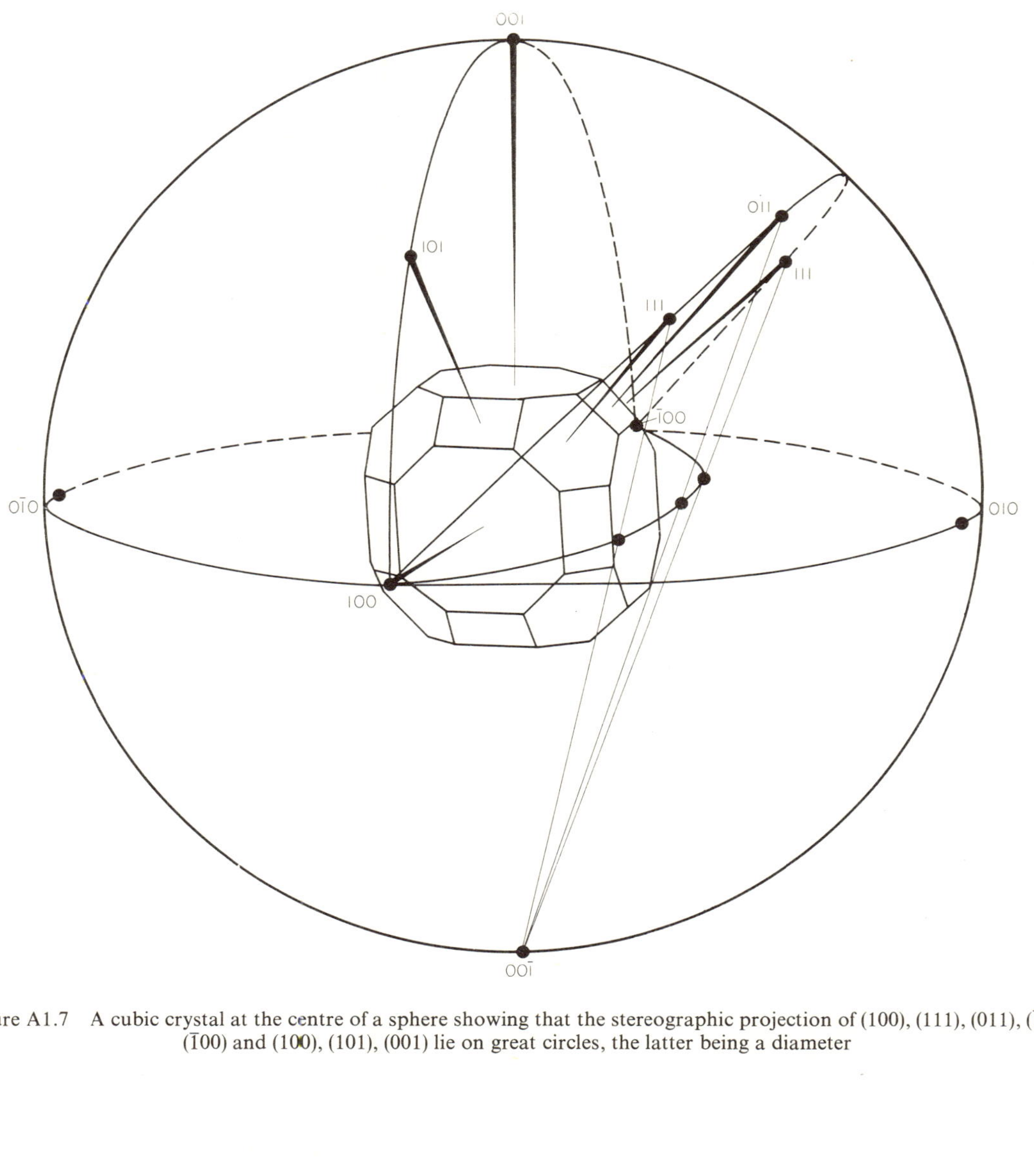

Figure A1.7 A cubic crystal at the centre of a sphere showing that the stereographic projection of (100), (111), (011), ($\bar{1}$11), ($\bar{1}$00) and (100), (101), (001) lie on great circles, the latter being a diameter

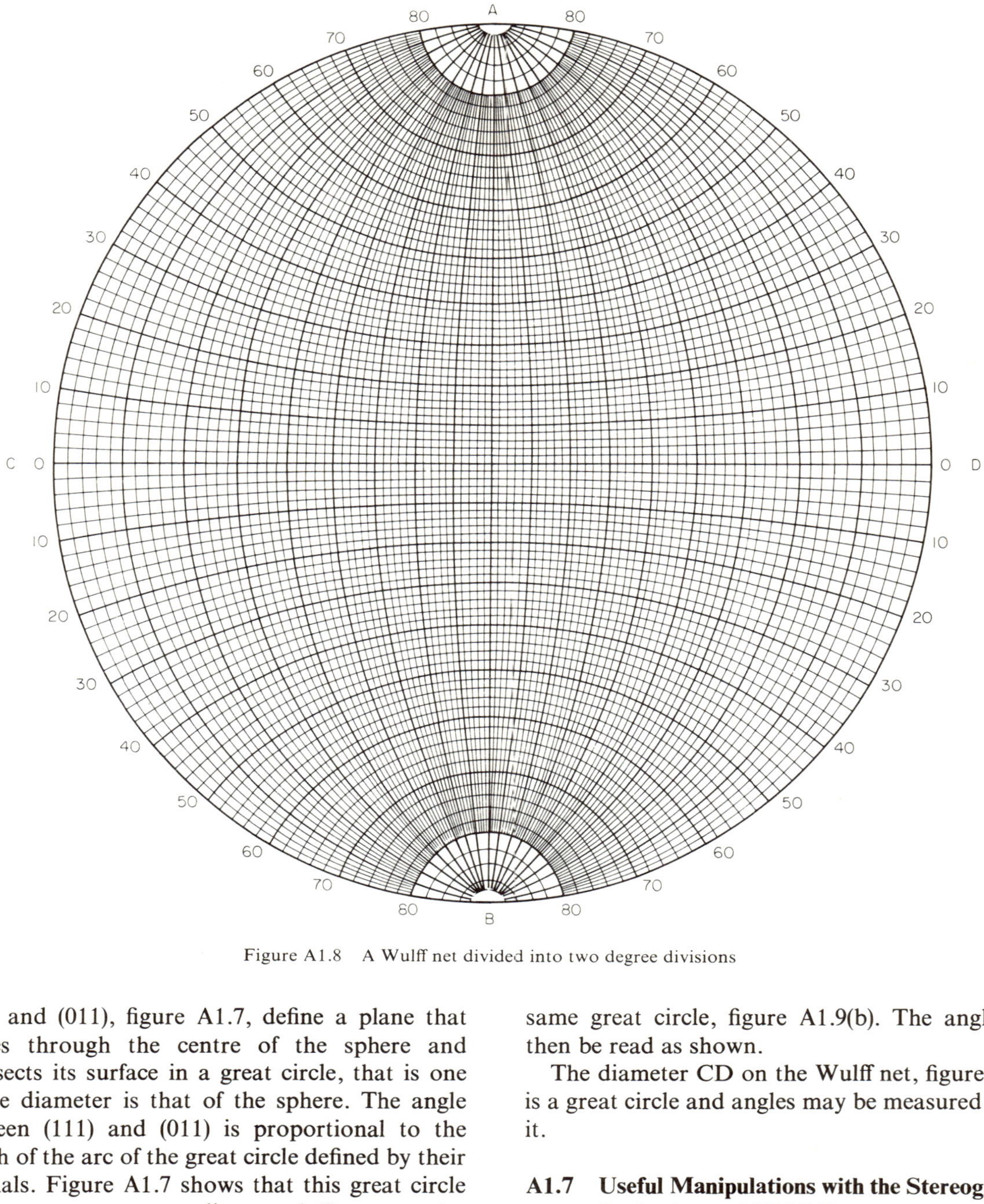

Figure A1.8 A Wulff net divided into two degree divisions

(111) and (011), figure A1.7, define a plane that passes through the centre of the sphere and intersects its surface in a great circle, that is one whose diameter is that of the sphere. The angle between (111) and (011) is proportional to the length of the arc of the great circle defined by their normals. Figure A1.7 shows that this great circle projects as an arc on a diameter.* Consequently, it is possible to measure the arc angle between (111) and (011) in terms of the distance along this arc between the plane normal projections. The device for doing this is a Wulff net, shown in figure A1.8. This consists of two sets of arcs. The first is an array of great circles on the same diameter AB. The second is a series of arcs centred on A and B such that their separation along any great circle corresponds to the same angle.† Thus to use the Wulff net to measure an angle between the projected plane normals P_1 and P_2, figure A1.9(a), the net is rotated about its centre until P_1 and P_2 lie on the same great circle, figure A1.9(b). The angle can then be read as shown.

The diameter CD on the Wulff net, figure A1.8, is a great circle and angles may be measured along it.

A1.7 Useful Manipulations with the Stereographic Projection and Wulff Net

In all of the following manipulations the Wulff net is used with its centre at the centre of the projection.

(1) To measure angles between any two planes or directions. This has been covered at the end of the previous section.

(2) To find the pole of a great circle. This is the projection of the axis of the zone of planes whose normals lie on the great circle. The Wulff net is aligned to superimpose on the great circle and the pole is constructed 90° along CD from the intersection of the great circle with CD, figure A1.10(a).

* Note that the great circle (100), (101), (001), figure A1.7, projects onto a diameter.

† Clearly the Wulff net corresponds to the projection of lines of latitude and longitude on the earth into the plane of the Greenwich meridian using the point 90° west on the equator as the projecting point.

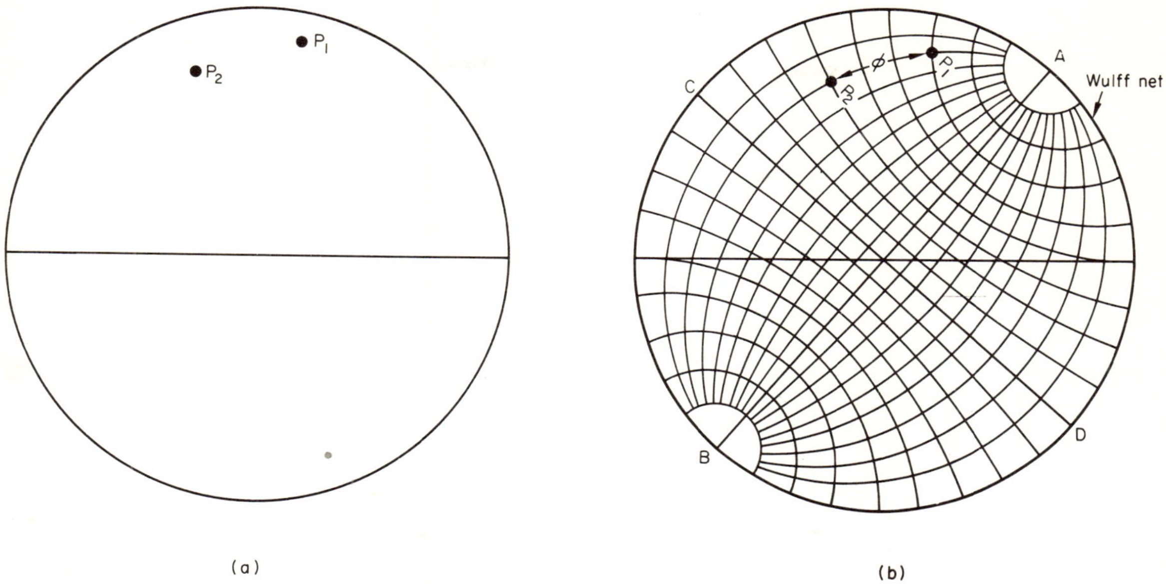

Figure A1.9 The measurement of the angle ϕ between poles P_1 and P_2 in the stereographic projection, using the Wulff net

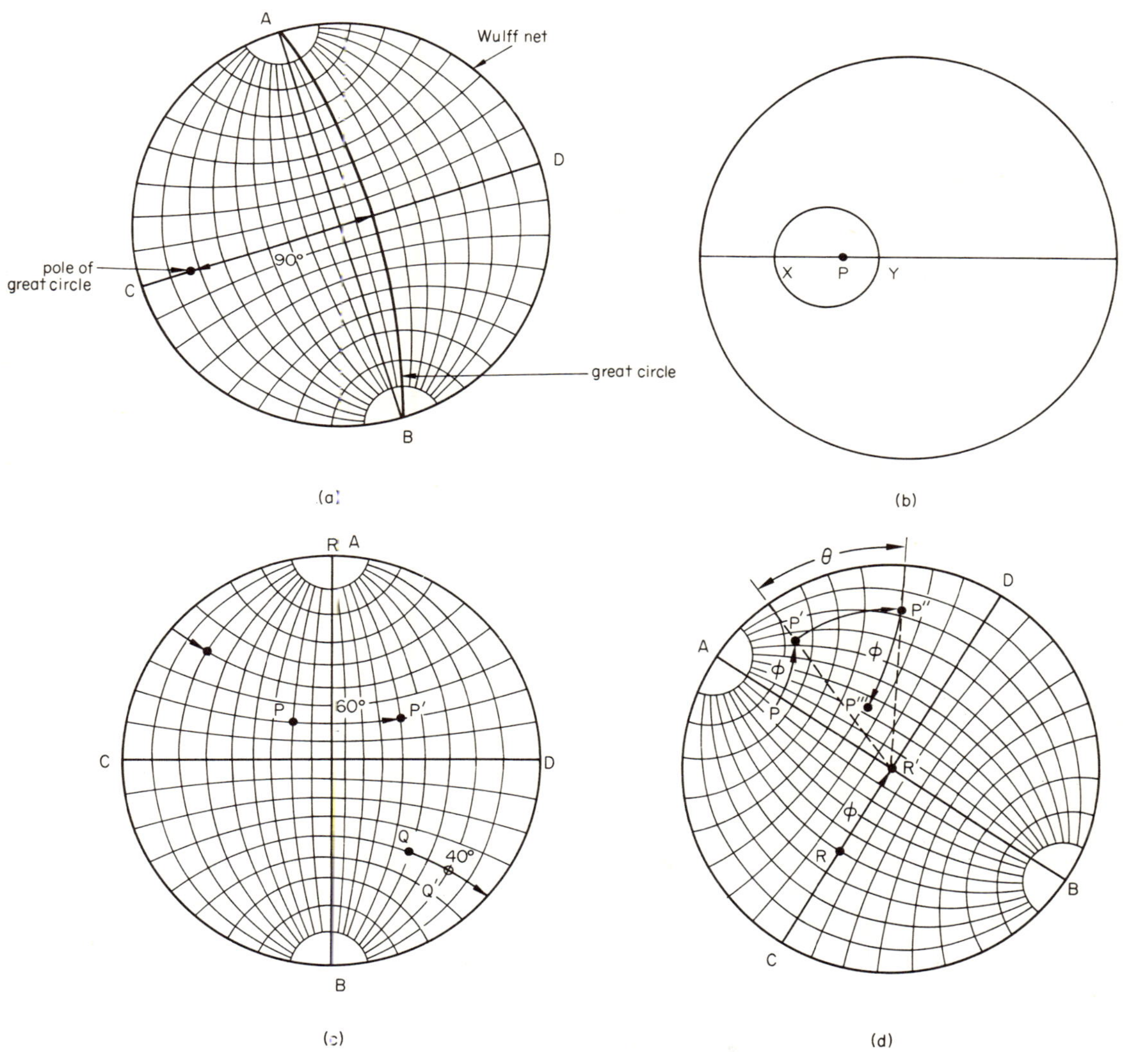

Figure A1.10 The use of the Wulff net (a) to find the pole of a great circle, (b) to construct a small circle about a pole P, (c) to rotate poles P and Q about a direction R on the edge of the stereogram, and (d) to rotate pole P about an axis R that does not lie on the edge of the stereogram

(a)

(b)

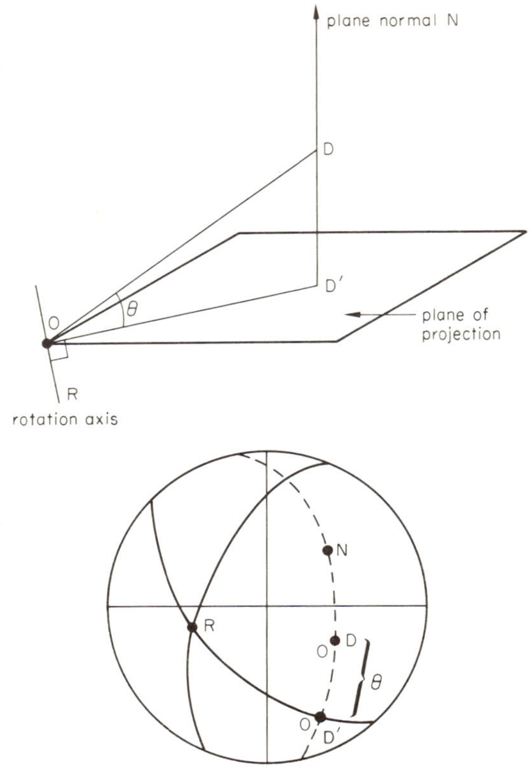

Figure A1.11 The projection of a particular crystal direction OD into a given plane (a) in real space, (b) on the stereographic projection

(3) To find the great circle corresponding to a pole. This is the reverse of manipulation (2).

(4) To construct a small circle about a pole. This corresponds to the projections of all plane normals at a given angle to the pole. Rotate the Wulff net so that either AB or CD cuts the pole. Measure the necessary angle from the pole P in figure A1.10(b) in both directions along either AB or CD to give the points X and Y. Bisect the line XY and draw the circle with XY as a diameter. Note that the geometric centre of this circle is not P.

(5) To rotate poles about an axis in the plane of projection, see figure A1.10(c). To rotate poles P and Q through 60°, in figure A1.10(c) about a pole R lying in the perimeter of the stereogram, set the A of the stereogram at R and measure along the arc shown the necessary amount to the positions P′ and Q′. Note that Q′ now must be regarded as projected from the north pole, see section A1.6, and is shown as an open circle, that is 'underneath' the stereogram.

(6) To rotate a pole through an angle θ about an axis R not on the perimeter of the stereogram. This procedure is shown in figure A1.10(d). Here R lies as shown and the net is rotated until CD cuts R. R and P are rotated by ϕ about AB as in (5) until R lies in the centre at R′ and P moves to P′. Then P′ is rotated to P″ by the required amount (θ) about R′; then R′ is rotated back to R about AB and P″ moves to P‴.

(7) To project a given direction OD into a specific plane, normal N, see figure A1.11(a). In three dimensions this involves rotating the direction OD down into the plane required, about an axis OR lying in the projection plane and perpendicular to OD, see figure A1.11(a). This is performed on the stereogram by drawing great circles corresponding to the plane normal N, and OD. These will intersect at R, a direction perpendicular to N and OD, see figure A1.11(b). Consideration of figure A1.11(a) will show that N, OD and OD′ are all perpendicular to R. Thus D′ always lies at the intersection of great circles with poles R and OD. The same result can be obtained by moving R to the centre then following the procedure outlined in (6).

A1.8 Useful Crystallographic Formulae for Various Crystal Structures

In interpreting electron diffraction patterns it is particularly useful to have available tables of interplanar spacings and angles between planes for the crystal structure of interest. These may be generated by computation from the formulae in table A1.1. Useful values of interplanar spacings and angles are listed in appendix 6 for the cubic and hexagonal crystal structures. The definitions of a, b, c, α, β, γ in table A1.1 are given for each crystal structure, illustrated in figure A1.1. Table A1.2 contains formulae for obtaining the indices of directions normal to planes and vice versa. Finally, table A1.3 lists crystal structures and lattice parameters of the elements which are crystalline at room temperature.

Appendix 1: Recommended Reading

Further information on crystallography and the use of the stereographic projection can be found in a number of books, including the following.

Gay, P. (1972). *The Crystalline State, An Introduction*, Oliver and Boyd, Edinburgh.

Johari, O., and Thomas, G. (1969). The stereographic projection and its applications. *Techniques of Metals Research* (ed. R. F. Bunshah), Interscience, New York.

Kelly, A., and Groves, G. W. (1970). *Crystallography and Crystal Defects*, Longmans, London.

Phillips, F. C. (1963). *An Introduction to Crystallography*, Longmans, London.

Smaill, J. S. (1972). *Metallurgical Stereographic Projections*, Hilger, London.

Appendix 1: References

Andrews, K. W., Dyson, D. J., and Keown, S. R. (1971). *Interpretation of Electron Diffraction Patterns*, Hilger, London.

Barrett, C. S., and Hassalski, T. B. (1968). *Structure of Metal and Alloys*, McGraw-Hill, New York.

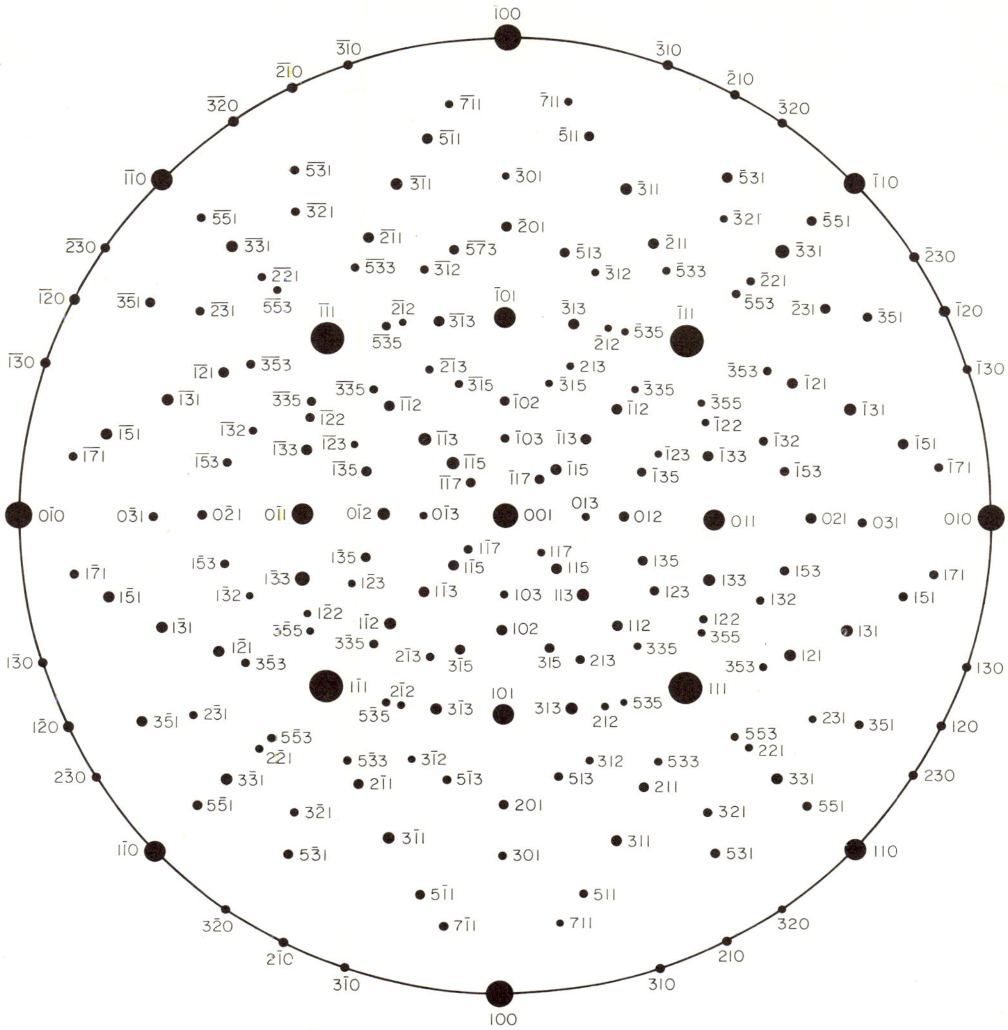

Figure A1.12 Standard stereographic projections for cubic structures: (a) (001); (b) (110); (c) (111); (d) (112)

Table A1.3 The lattice parameters and crystal structures of the elements crystalline at room temperature [After Barrett and Hassalski (1968)]

Element: form (transformation temp. °C)	Temp (°C)	Structure	Lattice constants a (Å)	Lattice constants b (Å)	c (Å) (α or β)
aluminum (Al)	25	f.c.c.	4.0496		
antimony (Sb)	25	rhomb.	4.5067		57°6′27″
	26	rhomb.	4.307	(hex. axes)	11.273
			z = 0.2335		
	4 °K	rhomb.	4.3007	z = 0.2336	11.222
arsenic (As)		rhomb.	4.131		α = 54°10′
barium (Ba)	R.T.	b.c.c.	5.019		
beryllium, α (Be)	20	c.p.h.	2.2856		3.5832
β? > 1250	1250	b.c.c.	2.55		
bismuth (Bi)	25	rhomb.	4.546		11.862
			(hex. axes) z = 0.2339		
	78 °K	rhomb.	4.535	z = 0.2341	11.814
	4 °K	rhomb.		z = 0.2340	11.862
cadmium (Cd)	21	c.p.h.	2.9788		5.6167
calcium, α (Ca)	18	f.c.c.	5.582		
γ 464 to m.p.	~500	b.c.c.	4.477		
carbon, diamond	20	cubic	3.5670		
graphite, α	20	hex.	2.4612		6.707

Table A1.3 (*continued*)

Element: form (transformation temp. °C)	Temp. (°C)	Structure	Lattice constants a (Å)	b (Å)	c (Å) (α or β)
carbon (*contd.*)					
graphite, β		rhomb.	2.4612		10.061
caesium (Cs)	−10	b.c.c.	6.14		
chromium (Cr)	20	b.c.c.	2.8846		
cobalt, α (Co)	18	c.p.h.	2.506		4.069
β stable ~450 to m.p.	18	f.c.c.	3.544		
copper (Cu)	20	f.c.c.	3.6147		
	0	f.c.c.	3.6029		
gallium (Ga)	20	orthorhomb.	4.5258	4.5198	7.6602
germanium (Ge)	25	cubic	5.6576		
gold (Au)	25	f.c.c.	4.0788		
hafnium α (Hf)	24	c.p.h.	3.1946		5.0511
indium (In)	R.T.	tetrag.	4.5979	(f.c. cell)	4.9467
	R.T.	tetrag.	3.2512	(b.c. cell)	4.9467
iodine (I)	26	orthorhomb.	4.79	7.25	9.78
iridium (Ir)	26	f.c.c.	3.8389		
iron, α (Fe)	20	b.c.c.	2.8664		
γ 911 to 1392	916	f.c.c.	3.6468		
δ 1392 to m.p.	1394	b.c.c.	2.9322		
lead (Pb)	25	f.c.c.	4.9502		
magnesium (Mg)	25	c.p.h.	3.2094		5.2105
manganese, α (Mn)	25	cubic	8.9139		
β 742 to 1095	25	cubic	6.315		
γ 1095 to 1133	1095	f.c.c.	3.862		
δ 1133 to m.p.	1134	b.c.c.	3.081		
molybdenum (Mo)	20	b.c.c.	3.1468		
nickel (Ni)	18	f.c.c.	3.5236		
niobium (Nb) (columbium)	20	b.c.c.	3.3007		
palladium (Pd)	22	f.c.c.	3.8907		
platinum (Pt)	20	f.c.c.	3.9239		
plutonium, α (Pu)	21	monoclin.	6.1835	4.8244 β = 101.81°	10.973
β 122 to 206	190	monoclin.	9.284	10.463	7.859 β = 92.13°
γ 206 to 319	235	orthorhomb.	3.159	5.768	10.162
δ 319 to 451	320	f.c.c.	4.637		
δ' 451 to 485	477	tetrag.	3.339		4.446
ε 476 to m.p.	490	b.c.c.	3.636		

APPENDIX 2. CRYSTALLOGRAPHIC TECHNIQUES FOR THE INTERPRETATION OF TRANSMISSION ELECTRON MICROGRAPHS OF MATERIALS WITH HEXAGONAL CRYSTAL STRUCTURE

A2.1 Introduction

The geometrical interpretation of transmission electron micrographs of hexagonal close-packed metals is more complicated than the equivalent interpretation of cubic metals for three reasons. Firstly, prominent zone axes are not in general normal to prominent planes, so a simple diffraction pattern commonly corresponds to a non-rational foil plane; secondly, there is no simple relationship between the plane ($hkil$) and the direction [$hkil$]; and, thirdly, the Miller–Bravais system of indexing directions in this crystal structure is not easy to visualise. The analysis of transmission electron micrographs obtained from hexagonal materials is based on a number of equations representing some important geometrical relationships between planes, directions, etc. These equations are first derived and then applied to particular problems.

A2.2 Crystallographic Relationships for the Hexagonal Lattice

As is well known, in the Miller–Bravais notation the hexagonal system is described by four axes, three of which are coplanar. The three coplanar axes, labelled a_1, a_2 and a_3, lie in the basal plane of the lattice and are 120° apart. The fourth axis is normal to this plane, and the right-hand rule applies for labelling the direction of the axes. The Miller–Bravais indices of a plane ($hkil$) are then the ratios of the reciprocals of the intercepts of the plane on the four axes (figure A2.1). If the intercepts on the axes a_1, a_2, a_3 are respectively a/h, a/k and $-a/i$ it follows from elementary geometry that

$$i = -(h + k) \tag{A2.1}$$

One of the three coplanar axes is therefore strictly unnecessary, and in fact is included only to demonstrate the symmetry of the crystal system. In some cases ($hkil$) indexing is written as ($hk{\cdot}l$) where $i = \cdot$.

However, the labelling of directions is less obvious. Again, as in the cartesian representation of a direction in figure A1.3, the direction is represented as a line joining the origin of the coordinate system to a point in space, and the direction indices [$uvtw$] are the indices of the end point of the line. However, u, v and t now represent successive displacements parallel to a_1, a_2 and a_3 and are chosen so that equality (A2.1) is satisfied. This produces a representation which is now non-cartesian as figure A2.2 demonstrates, and the t index is no longer a dummy. To obtain the equivalent components of the direction in the three-index system related to the non-coplanar axes a_1, a_2 and c, where now the indices can be regarded as components of a vector in a skew three-space, we now set the unit vector along the axes as a, a and c, where a and c are the lattice spacings for the appropriate metal and c/a is the axial ratio. The magnitude of direction OR in the basal plane (figure A2.2) can be found by taking the square root of the scalar product of the vector

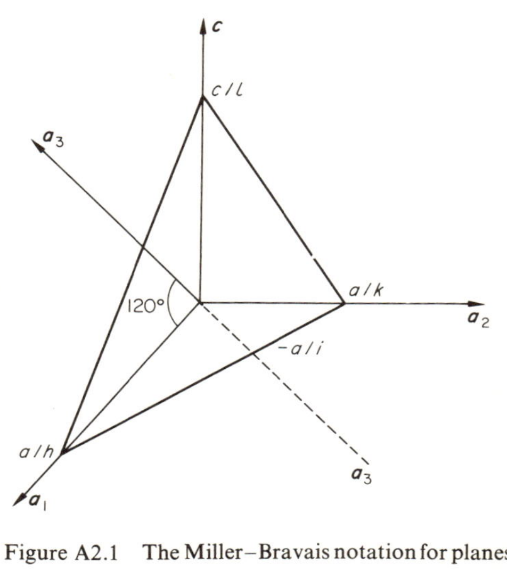

Figure A2.1 The Miller–Bravais notation for planes

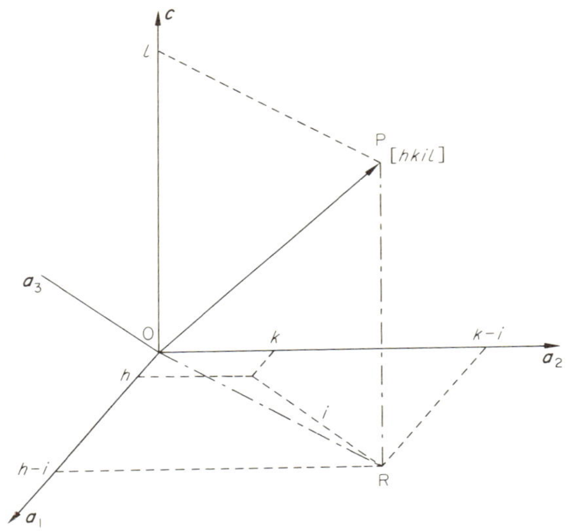

Figure A2.2 The Miller–Bravais notation for directions

specifying OR with itself and is given by

$$OR = \{3a^2(u^2 + v^2 + uv)\}^{1/2}$$

where a is the interatomic spacing in the basal plane. A distance OP (figure A2.2) that has a component out of the basal plane has a magnitude

$$OP = \{3a^2(u^2 + v^2 + uv) + c^2w^2\}^{1/2} \quad \text{(A2.2)}$$

A2.2.1 *Angles between Two Directions, ϕ*

The angle between two crystallographic directions is found by means of the cosine law

$$Z^2 = X^2 + Y^2 - 2XY \cos \phi$$

where X, Y and Z are magnitudes of vectors and ϕ is the angle between X and Y, as seen in figure A2.3.

Directions X and Y are known and Z can be found.* The magnitudes can be determined by the formula for distance. The cosine of the angle between X and Y is given by

$$\cos \phi = \frac{Dd + Ee + \frac{1}{2}(De + Ed) + \frac{1}{3}Gg(c/a)^2}{\{D^2 + E^2 + DE + (G^2/3)(c/a)^2\}^{1/2} \times \{d^2 + e^2 + de + (g^2/3)(c/a)^2\}^{1/2}} \quad \text{(A2.3)}$$

where $X = [DEFG]$ and $Y = [defg]$.

A2.2.2 *Indices [defg] of the Normal to the Plane (hkil)*

The direction $[d, e, f, g]$ perpendicular to plane (h, k, i, l) can be found by forming the scalar product of the vector representing d, e, f, g with two vectors in (h, k, i, l) (figure A2.4).

Direction $[d, e, f, g]$ intersects (h, k, i, l) at a point which is some multiple n times ad, ae, af and cg. Since the intercepts of the plane with the axes are known, vectors **AX**, **BX** and **CX** can be found, for example

$$\mathbf{AX} = (2nda + nea - a/h)\boldsymbol{a}_1 + (2nea + nda)\boldsymbol{a}_2 + ngc\,\mathbf{c}$$

The direction d, e, f, g will have components such that

$$\mathbf{OX} = (2nda + nea)\boldsymbol{a}_1 + (2nea + nda)\boldsymbol{a}_2 + ngc\,\mathbf{c}$$

Forming the scalar product of **AX** and **OX** and setting it equal to zero, since the two vectors are perpendicular to each other, gives

$$\mathbf{OX} \cdot \mathbf{AX} = 4(nda) + 4(nea) + \tfrac{11}{2}(nda)(nea) - \frac{3}{2}\frac{(nda)a}{h} + (ngc)^2 = 0$$

Analogous equations can be formed for **BX** and **CX**.

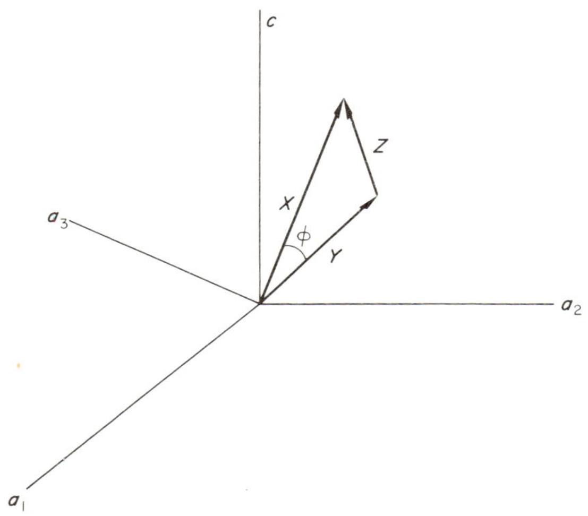

Figure A2.3 Angles between directions

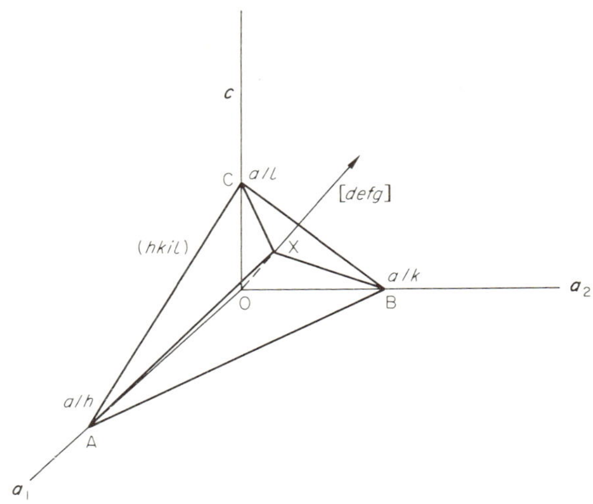

Figure A2.4 Indices of a plane normal

* From equation (A2.1).

One of the indices, d, e, f, g, may be arbitrarily chosen. This is usually done in such a way as to make the indices the smallest integer values. The index d will be set equal to h.

The normal to the plane (h, k, i, l) will have indices

$$[defg] = h, k, i, \tfrac{3}{2}(a/c)^2 l \qquad \text{(A2.4)}$$

A2.2.3 *Directions* $[wxyz]$ *Lying in a Plane* $(hkil)$

This expression is easily calculated by taking the scalar product of $[wxyz]$ with the plane normal and setting equal to zero.

The condition becomes

$$uh + vk + ti + wl = 0 \qquad \text{(A2.5)}$$

This expression can also be used to calculate the planes containing a given direction. While it seems that relations (A2.4) and (A2.5) appear to require knowledge of each other, this is not in fact so. Three directions lying in a plane can immediately be constructed from the knowledge of the intercepts made by the plane on the axes.

A2.2.4 *Angle* ϕ *between Two Planes*

The angle between two planes is the same as the angle between their normals, so combining (A2.3) and (A2.4) the cosine of the angle ϕ between two planes $(hkil)$ and $(defg)$ is given by

$$\cos\phi = \frac{hd + ke + \frac{1}{2}(he + kd) + \frac{3}{4}lg(a/c)^2}{\{h^2 + k^2 + hk + \frac{3}{4}l^2(a/c)^2\}^{1/2} \times \{d^2 + e^2 + de + \frac{3}{4}g^2(a/c)^2\}^{1/2}} \qquad \text{(A2.6)}$$

This is identical with the expression (calculated by using different techniques) given in standard crystallographic texts. The angle between a direction and a plane can similarly be calculated by using (A2.3) and (A2.4).

A2.2.5 *Direction of the Intersection of Two Planes*

This is easily calculated by applying (A2.5) to both planes: the direction $[wxyz]$ which satisfies (A2.5) for both planes is plainly the direction of their line of intersection.

A2.3 Stereographic Manipulations in the Hexagonal Lattice

A number of geometrical calculations such as those outlined above can be performed on a double stereogram on which are represented both the poles of planes and crystallographic directions. Of course, it is necessary to orient the plane and direction projections relative to one another, and in the hexagonal system since [0001] is normal to (0001) and $[hki0]$ is normal to $(hki0)$, it is easiest to use a standard basal projection with the direction and plane indices coincident at the centre and the rim. In effect such a stereogram is a chart for transforming from one index system to another and, viewed in this light, is quite general. The concept of the double stereogram can be applied to any crystal system. A general point in such a stereogram may be regarded as a plane (properly, the projection of the pole of the plane) or as a direction and indexed accordingly. It follows therefore that the projection of a plane and the direction normal to it coincide, as usual, but they have different indices.

To determine Burgers' vectors, slip planes, and the geometry of dislocation interactions, it is necessary to perform a small number of simple operations using the double stereogram. In general geometrical operations are performed in the foil plane, which is determined from a diffraction pattern as described below.

A2.3.1 *Indexing Diffraction Patterns*

As pointed out in section 2.2.2.1, each spot in the diffraction pattern corresponds to a set of planes, almost parallel to the electron beam. The spots are indexed $(hkil)$ using the procedure in section 2.7.2, but substituting relation (A2.6) for (2.22). Vector addition may be used to simplify indexing of the complete pattern, once the initial indexing of two spots has been accomplished. The zone axis $|uvtw|$ of the pattern may be obtained from the relations

$$\left.\begin{aligned} u &= l_2(2k_1 + h_1) - l_1(2k_2 + h_2) \\ v &= l_1(2h_2 + k_1) - l_2(2h_1 + k_1) \\ w &= 3(h_1k_2 - h_2k_1) \\ t &= -(u + v) \end{aligned}\right\} \qquad \text{(A2.7)}$$

where the indexed spot $(h_2k_2i_2l_2)$ is positioned anticlockwise relative to $(h_1k_1i_1l_1)$. As an alternative to relations (A2.7), relation (2.23a) may be used with the Miller $(hkil)$ indices to give notional values for u', v', w' which may be converted to the correct u, v, t, w by the relations

$$\left.\begin{aligned} u &= \tfrac{1}{3}(2u' - v') \\ v &= \tfrac{1}{3}(2v' - u') \\ t &= -\tfrac{1}{3}(v' + u') \\ w &= w' \end{aligned}\right\} \qquad \text{(A2.8)}$$

All spots in the pattern must satisfy relation (A2.5).

Having determined the beam direction $\boldsymbol{B}$, it may be required to determine the foil plane (if zero specimen tilt). Unlike the cubic case, the plane normal to $\boldsymbol{B}$ does not have the same indices as $\boldsymbol{B}$.

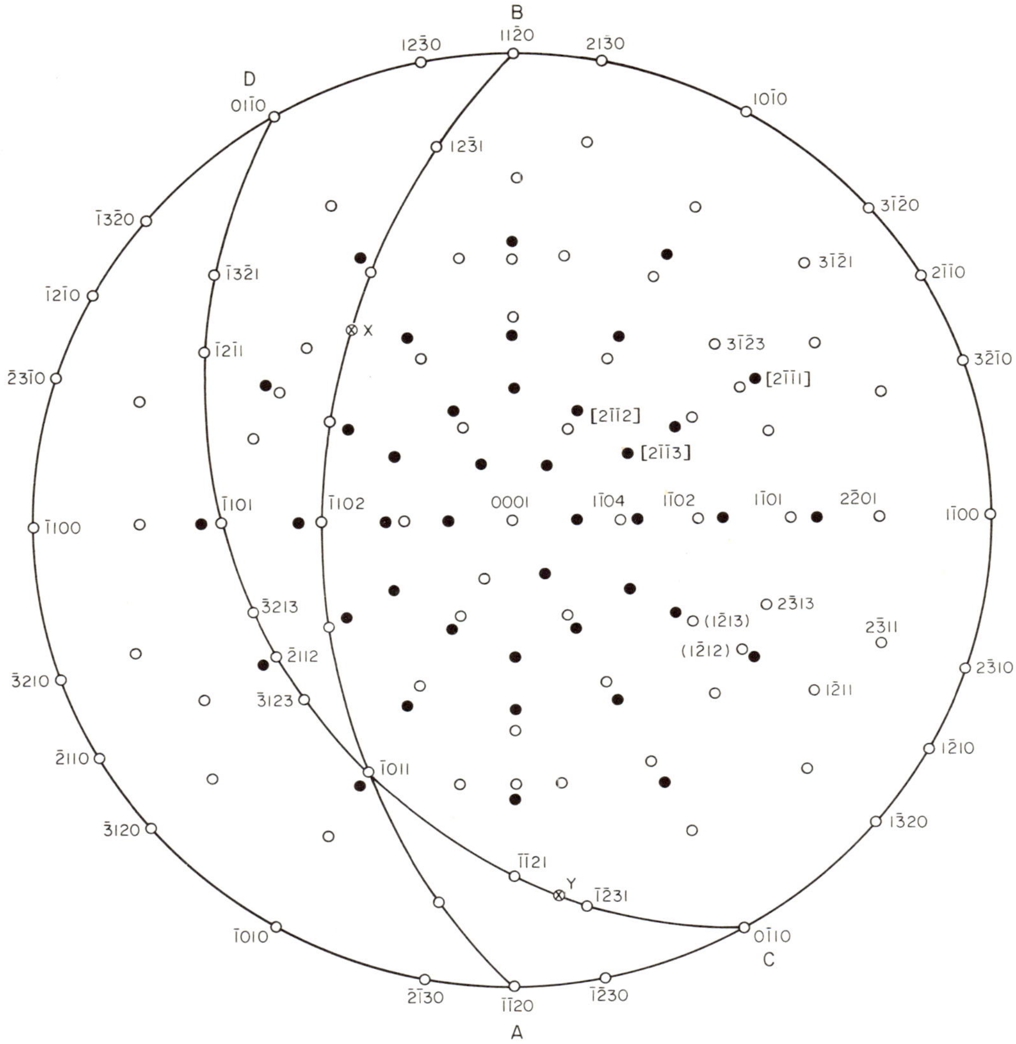

Figure A2.5 A double stereogram for axial ratio: $c/a = 1.62$; ○ planes; ● directions. For all *hki*0, and 0001, directions and planes superimpose. These are shown as open circles for clarity

The indices of the plane normal to ***B*** are given by

$$\left.\begin{aligned} h &= u \\ k &= v \\ l &= \tfrac{2}{3}(c/a)^2 w \\ i &= -(h + k) \end{aligned}\right\} \qquad \text{(A2.9)}$$

and

The zone of reflecting planes may be represented on the stereogram by a great circle passing through the poles corresponding to the planes giving rise to the reflections. A typical example is shown as the great circle AB in figure A2.5. The zone axis is then the direction which lies in all the planes, the pole of the great circle. In this case, the zone axis is [$1\bar{1}01$], and the foil plane is the plane normal to this. As can be seen, this foil plane has non-rational indices: it is fairly close to ($4\bar{4}07$); using relation (A2.9) the indices may be shown to be ($1, \bar{1}, 0, 1.679$). The use of the double stereogram will be demonstrated in section A2.3.4 by using this foil plane as an example.

A2.3.2 *Planes Containing a Given Direction*

A contrast experiment performed on a single dislocation gives a Burgers vector, but no slip plane. It is frequently instructive to know which planes contain this direction. For example, dislocations having a Burgers vector parallel to the prism face diagonals have occasionally been reported: these directions have indices $\langle 11\bar{2}3\rangle$. For the particular case of directions [$2\bar{1}\bar{1}3$], the expression (A2.5) indicates that planes (*hkil*) contain this direction, where

$$h + l = 0 \qquad \text{(A2.10)}$$

Since [$2\bar{1}\bar{1}3$] lies in all those planes whose poles are normal to it, the locus of these planes is the great circle drawn with $[2\bar{1}\bar{1}3]$ as a pole. This is shown as CD in figure A2.5, and it can be seen that in fact the indices of the planes satisfy relation (A2.10). As an extexsion of this, the directiox of the intersection of two planes is the pole of the great circle drawn through the two planes.

A2.3.3 *Contrast Experiments*

It is well known, see section 3.4.7, that the Burgers vector $\boldsymbol{b}$ of dislocations can be determined by tilting the foil to produce images under two-beam conditions with various values of $\boldsymbol{g} \cdot \boldsymbol{b}$, where $\boldsymbol{g}$ is the operative reflection vector, see section 2.4. If isotropic elasticity can be applied, when $\boldsymbol{g} \cdot \boldsymbol{b} = 0$ contrast disappears. However, if the material is elastically anisotropic, contrast will not disappear and, as described in section 3, image matching techniques must be used to determine the Burgers vector of dislocations. Nevertheless it is often useful to estimate quickly from a stereogram under which reflecting conditions $\boldsymbol{g} \cdot \boldsymbol{b} = 0$ to assess what tilting and contrast experiments are worth-while on a particular specimen. Geometrically, $\boldsymbol{g} \cdot \boldsymbol{b} = 0$ is equivalent to the statement that the direction of the Burgers vector must lie in the plane responsible for the operating reflection. Consider the case of a foil giving the diffraction pattern represented by the great circle AB (figure A2.5). Plainly, the $(\bar{1}011)$ plane contains the $[2\bar{1}\bar{1}3]$ direction, and thus any dislocations having a Burgers vector parallel to this will go out of contrast when the foil is tilted so that this reflection is the only one operating. Similarly, the $(\bar{1}102)$ plane contains the $[11\bar{2}0]$ direction, and dislocations for which $\boldsymbol{b} = (a/3)\,[11\bar{2}0]$ will be out of contrast for this operating reflection. Conversely, if a set of dislocations are observed to go out of contrast for a particular reflection, the Burgers vector must lie parallel to a direction lying on the great circle drawn with the reflecting plane as a pole.

A2.3.4 *Dislocation Geometry—Projection of Directions*

In some cases, even if contrast experiments cannot be performed, some indication of the Burgers vector can be obtained from the geometry of the dislocation arrangement. In many instances, dislocations tend to be either edge or screw and thus lie normal to or parallel to the Burgers vector in the slip plane. Consider, for example, slip on the $(\bar{1}101)$ pyramidal plane. This plane contains three possible Burgers' vectors, $(a/3)\,[11\bar{2}0]$, $\{(c^2 + a^2)^{1/2}/3\}\,[2\bar{1}\bar{1}3]$ and $\{(c^2 + a^2)^{1/2}/3\}\,[1\bar{2}13]$, as may be determined by drawing the great circle with $(\bar{1}101)$ as a pole. Geometrically, the projection of a direction into the foil plane is the trace of a plane normal to the foil plane containing the required direction: thus, the projection of $[2\bar{1}\bar{1}3]$ in $(4\bar{4}07)$ is the trace of $(\bar{1}011)$. The direction of this trace is the pole of the great circle passing through $(\bar{1}101)$ and $(4\bar{4}07)$, shown as X in figure A2.5. The angle between the direction and its projection (which gives its apparent length in the foil plane) is the complement of the angle between the directions and the normal to the foil plane, in this case $[1\bar{1}01]$. For $[2\bar{1}\bar{1}3]$ this angle is $90 - 24 = 66°$. The direction in $(\bar{1}101)$ normal to $[2\bar{1}\bar{1}3]$ is the pole of the great circle drawn through these two points, labelled Y in figure A2.5. This may be projected into the foil plane in exactly the same way. It is thus easy to prepare a map of projected edge and screw directions and projected lengths for the possible slip systems, which may then be compared with the dislocation images. In general, this is rather easier than attempting to determine the direction of the dislocation line directly, although this can be done by exactly the same method.

A2.4 Crystallographic Data for the Hexagonal Lattice

To plot a double stereogram for a hexagonal material it is necessary to know both the angles between planes and the angles between directions for the given c/a ratio. These values have been tabulated (Rarey *et al.*, 1966) and computer programmes are available which produce values (Johari and Thomas, 1970) and plot stereograms (Metzbower, 1969). Formulae for crystallographic relationships have also been given by Otte and Crocker (1965, 1966).

Appendix 2: References

Johari, O., and Thomas, G. (1970). The stereographic projection and its applications. *Techniques in Metals Research* (ed. R. F. Bunsah), vol. 2C, Wiley, New York.

Metzbower, E. A. (1969). *Trans. A.I.M.E.*, **245**, 435.

Otte, H. M., and Crocker, A. G. (1965). *Phys. Stat. Sol.*, **9**, 441.

—— (1966). *Phys. Stat. Sol.*, **16**, K25.

Rarey, C. R., Stringer, J., and Edington, J. W. (1966). *Trans. A.I.M.E.*, **236**, 811.

APPENDIX 4. STANDARD SPOT PATTERNS

This appendix includes diagrams of standard spot diffraction patterns for both cubic and hexagonal crystal structures. In the cubic cases each pattern for a given zone axis (z, defined in appendix 1) has the same, six-, four-, three- or two-fold symmetry but the reflections that occur obey the rules outlined in table 2.1. The patterns for the hexagonal crystal structure are indexed using the Miller–Bravais system outlined in appendix 2, assuming exact close packing, that is an axial ratio (c/a) of 1.633. The positions of the spots will change when the c/a changes. It is recommended that standard patterns be constructed for the c/a value corresponding to the actual material studied using the tables of Rarey, Stringer and Edington (1966). In figures A4.1–A4.4, the indexing procedure outlined in section 2.7.2.1 has been followed.

The positions of superlattice reflections are also shown in figures A4.1 and A4.2. The intensities of the spots will depend upon their structure factor as described in section 2.3.1. The reflections that occur in cubic crystal structures are shown in tables A4.1 and A4.2.

Appendix 4: Reference

Rarey, C. R., Stringer, J., and Edington, J. W. (1966). *Trans. A.I.M.E.*, **236**, 811.

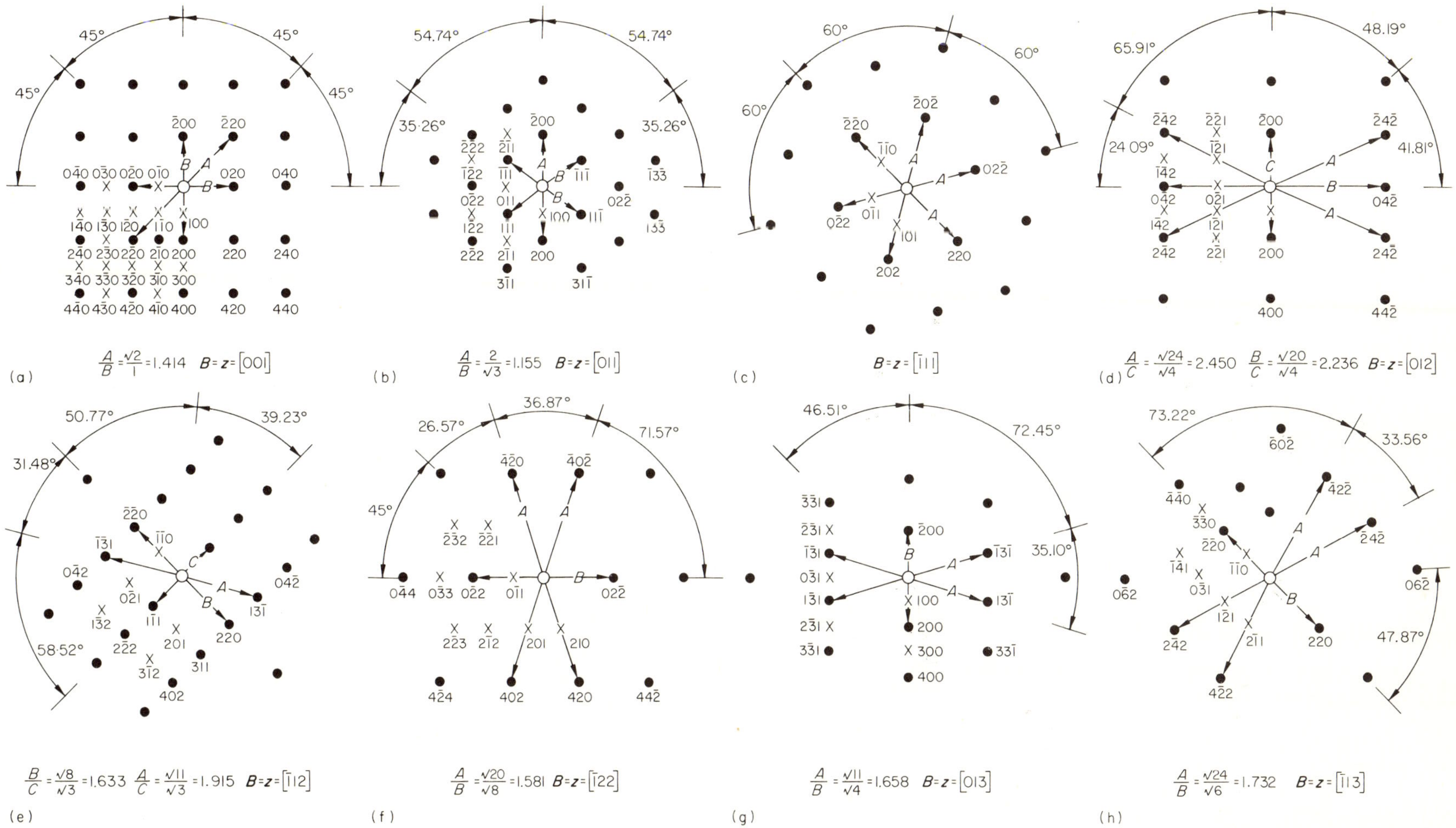

Figure A4.1 Single-crystal spot transmission electron diffraction patterns for the f.c.c. crystal structure ($u^2 + v^2 + w^2 < 22$). The zone axis z, defined in appendix 1, is the beam direction $\boldsymbol{B}$ defined in section 2.7.2, as indicated. The crosses in one quadrant of the diagram indicate the positions of the spots for the ordered f.c.c. ($L1_2$) unit cell. The complete pattern may be generated by repeating these spots in the remainder of the pattern and indexing using the addition of vectors, see section 2.7.2.1

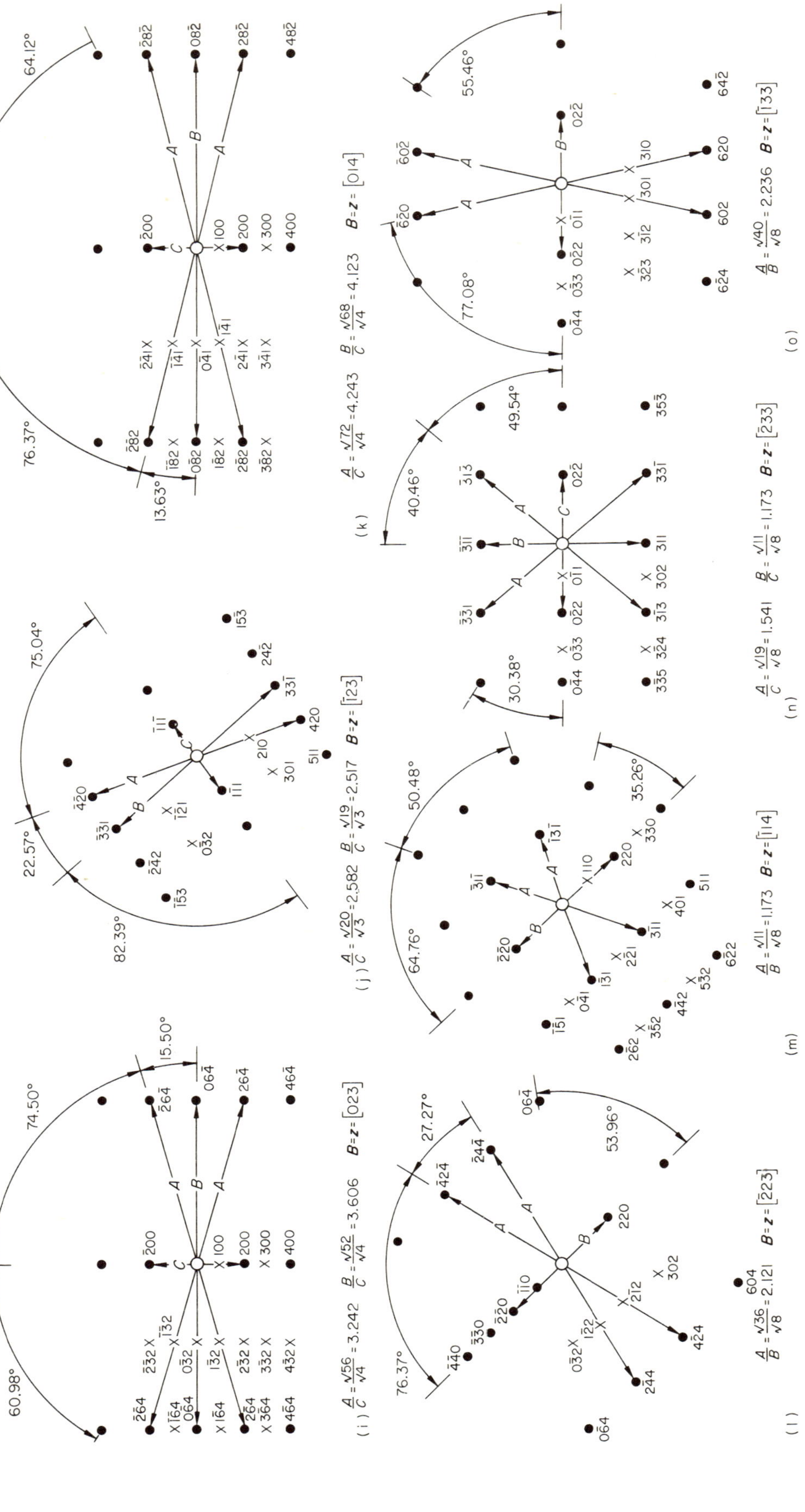

Figure A4.1 *(continued)*

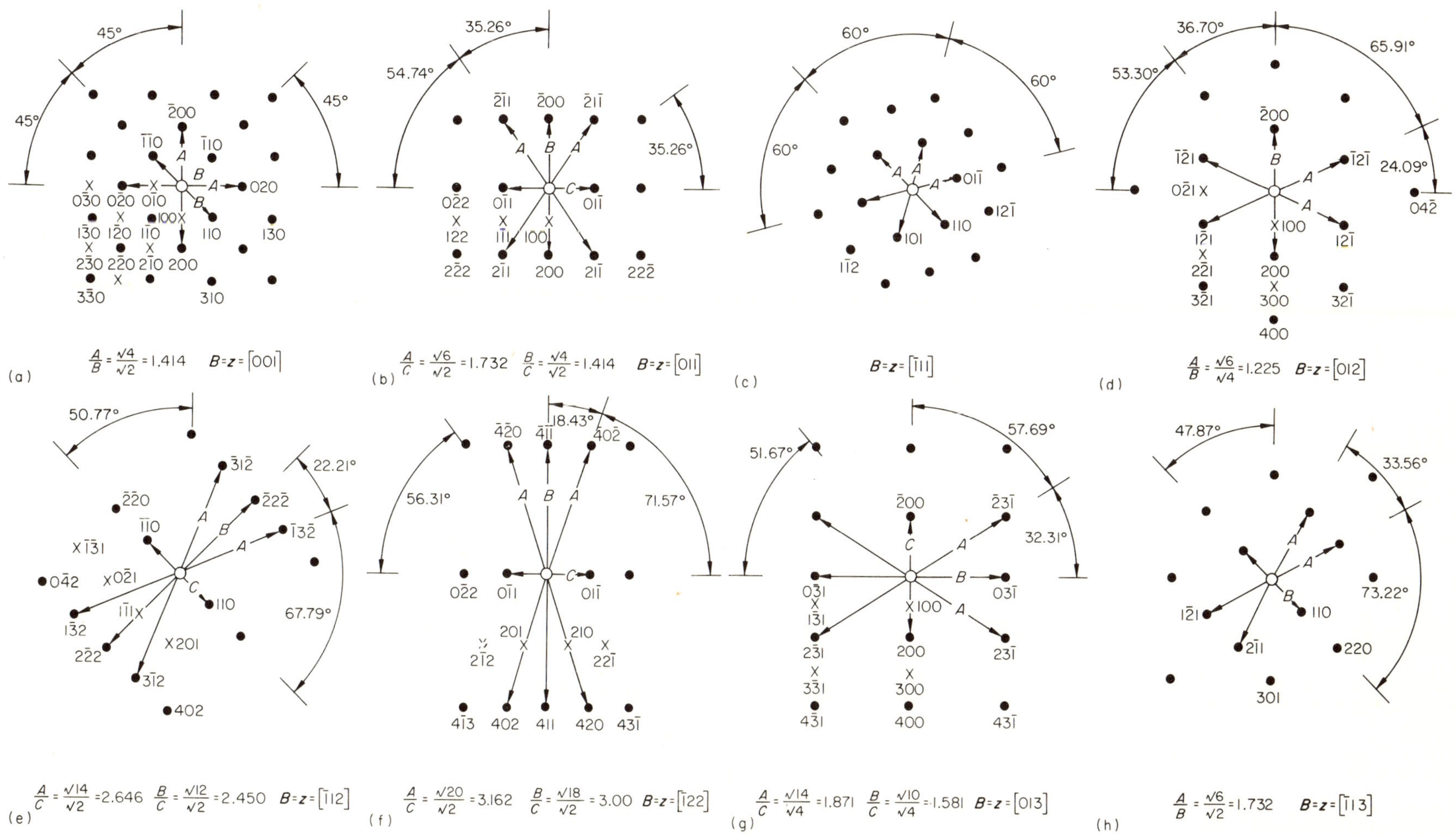

Figure A4.2 Single-crystal spot transmission electron diffraction patterns for the b.c.c. crystal structure ($u^2 + v^2 + w^2 \leqslant 22$). The zone axis ***z***, defined in appendix 1, is the beam direction ***B***, defined in section 2.7.2, as indicated. The crosses in one quadrant of the diagram indicate the positions of the spots for the ordered b.c.c. (B_2) unit cell. The complete pattern may be generated by repeating these spots in the remainder of the pattern and indexing using the addition of vectors, see section 2.7.2.1

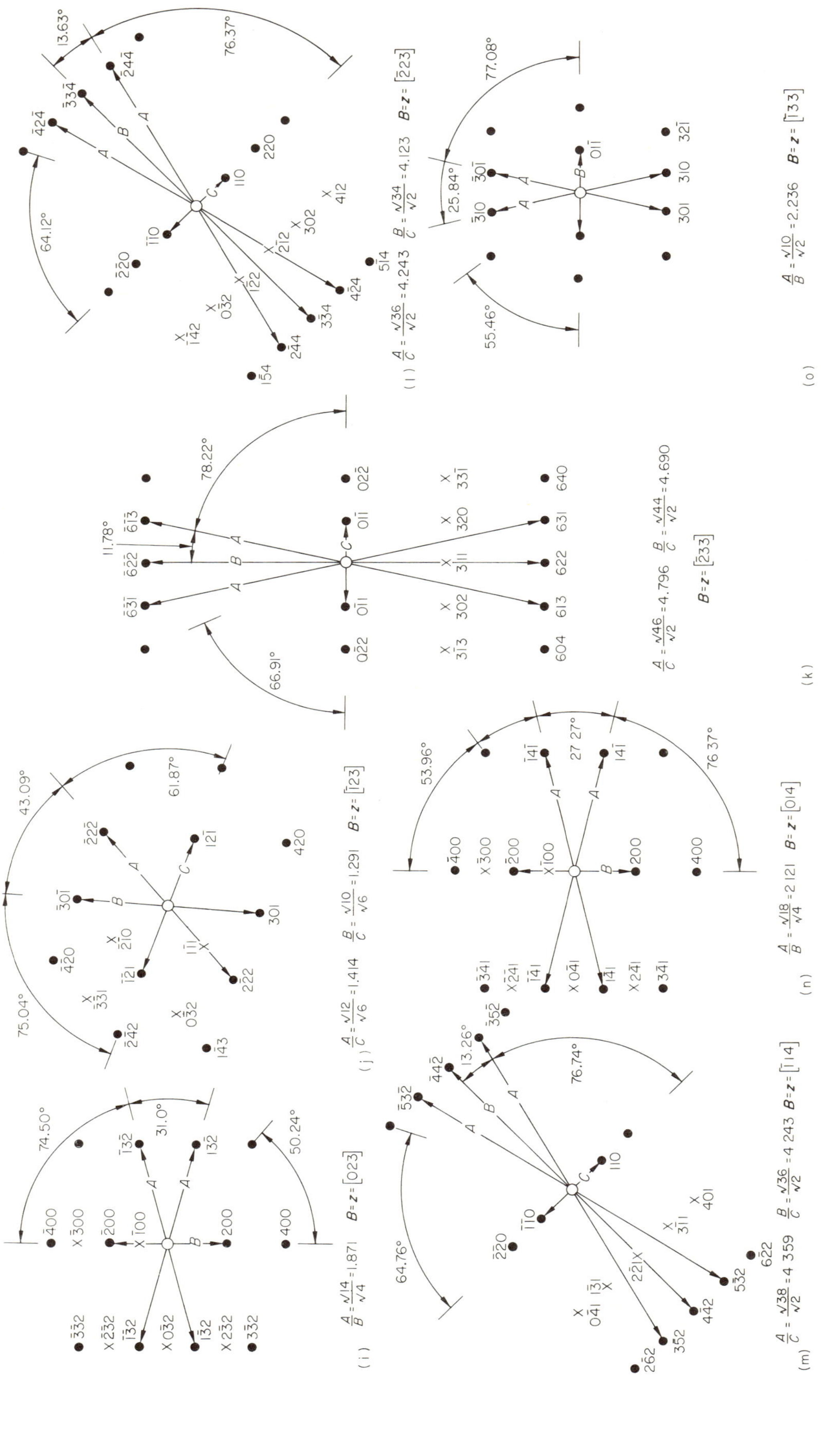

Figure A4.2 *(continued)*

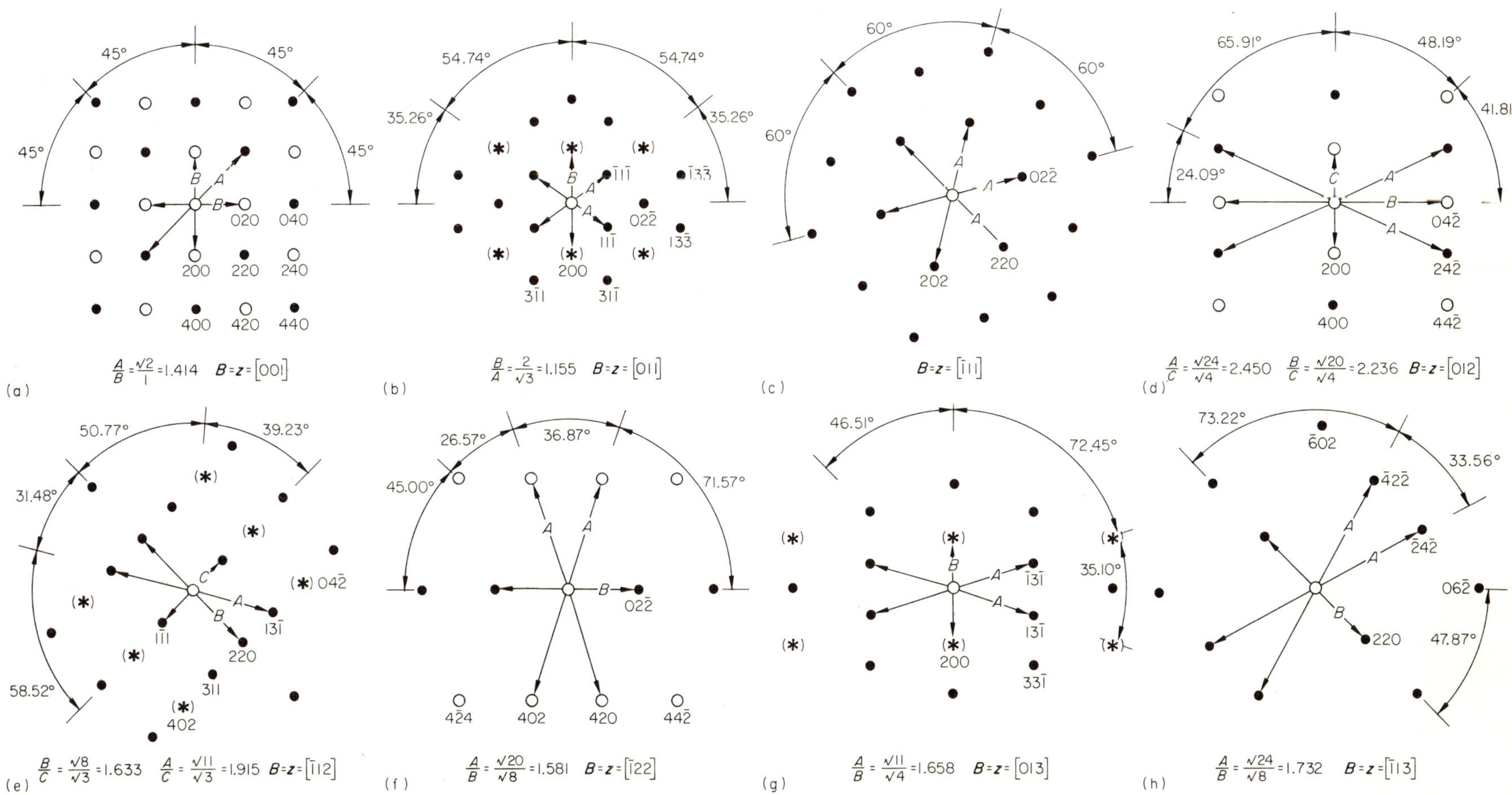

Figure A4.3 Single-crystal spot transmission electron diffraction patterns for diamond structures ($u^2 + v^2 + w^2 \leqslant 22$): ○ additional weaker spots; * spots that could arise from double diffraction. The zone axis z, defined in appendix 1, is the beam direction $\boldsymbol{B}$, defined in section 2.7.2, as indicated [After Andrews *et al.* (1971)]

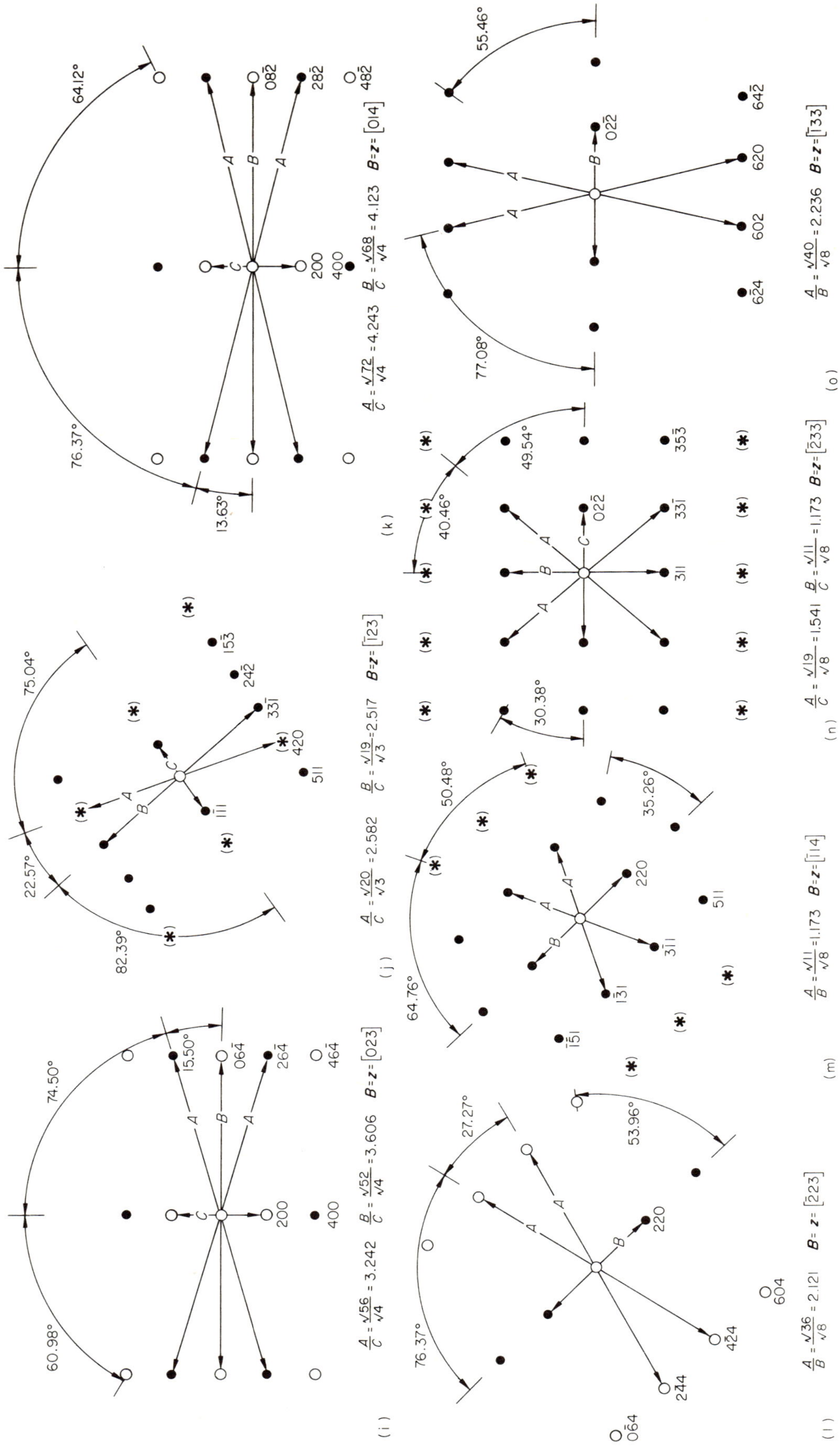

Figure A4.3 *(continued)*

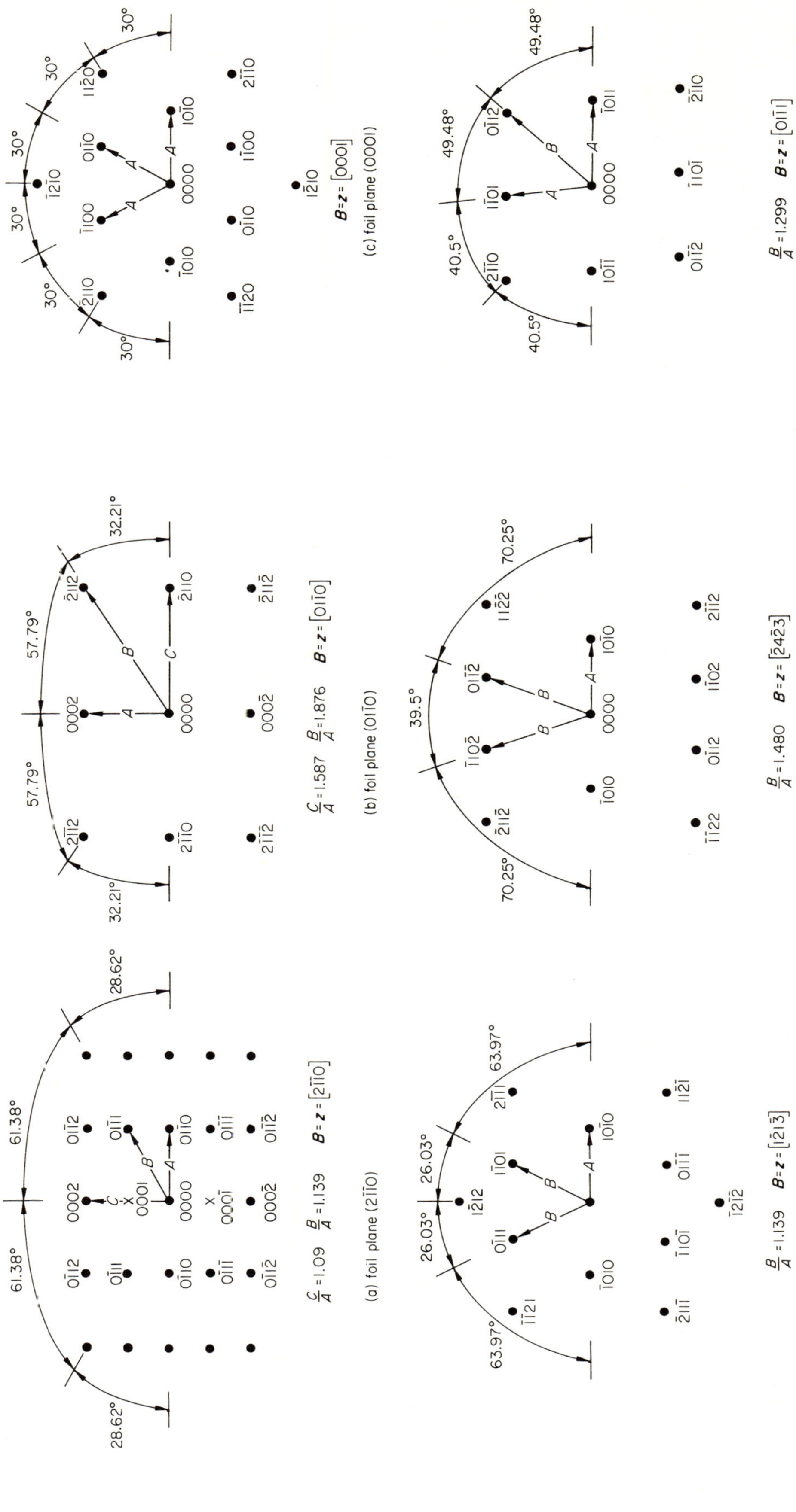

Figure A4.4 Single-crystal spot transmission electron diffraction patterns for the c.p.h. crystal structure. The zone axis, $\boldsymbol{z}$, defined in appendix 1, is the beam direction $\boldsymbol{B}$, defined in section 2.7.2. The crosses indicate the positions of reflections forbidden by the structure factor but occurring in the pattern by double diffraction

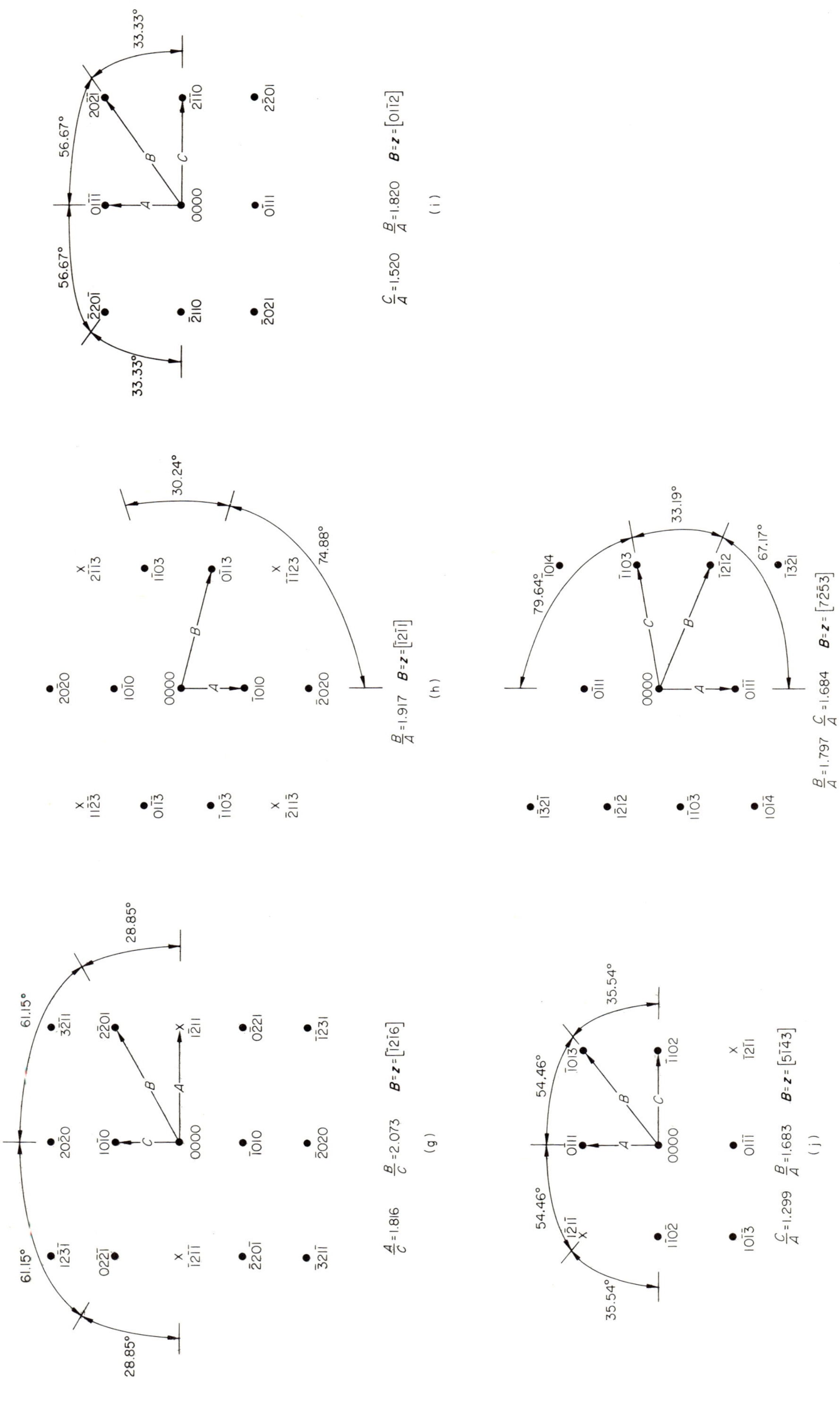

Figure A4.4 *(continued)*

Table A4.1 Occurrence of reflections for the cubic crystal structures

Line no. $N = h^2+k^2+l^2$	hkl indices	$N^{1/2} = (h^2+k^2+l^2)^{1/2}$		f.c.c.	diamond	Line no. $N = h^2+k^2+l^2$	hkl indices	$N^{1/2} = (h^2+k^2+l^2)^{1/2}$	b.c.c.	f.c.c.	diamond
1	100	1.00				33	522, 441	5.745			
2	110	1.414	×			34	530, 433	5.831	×		
3	111	1.732		×	×	35	531	5.916		×	×
4	200	2.00	×	×	×	36	600, 442	6.00	×	×	×
5	210	2.236				37	610	6.083			
6	211	2.450	×			38	611, 532	6.164	×		
7	—	—				39	—	—			
8	220	2.828	×	×	×	40	620	6.325	×	×	×
9	300, 221	3.00				41	621, 540, 443	6.403			
10	310	3.162	×			42	541	6.481	×		
11	311	3.317		×	×	43	533	6.557		×	×
12	222	3.464	×	×		44	622	6.633	×	×	×
13	320	3.606				45	630, 542	6.708			
14	321	3.742	×			46	631	6.782	×		
15	—	—				47	—	—			
16	400	4.00	×	×	×	48	444	6.928	×	×	
17	410, 322	4.123				49	700, 632	7.00			
18	411, 330	4.243	×			50	710, 550, 543	7.071	×		
19	331	4.359		×	×	51	711, 551	7.141		×	×
20	420	4.472	×	×	×	52	640	7.211	×	×	
21	421	4.583				53	720, 641	7.280			
22	332	4.690	×			54	721, 633, 552	7.349	×		
23	—	—				55	—	—			
24	422	4.899	×	×	×	56	642	7.483	×	×	×
25	500, 430	5.00				57	722, 544	7.550			
26	510, 431	5.099	×			58	730	7.616	×		
27	511, 333	5.196		×	×	59	731, 553	7.681		×	×
28	—	—				60	—	—			
29	520, 432	5.385				61	650, 643	7.810			
30	521	5.477	×			62	732, 651	7.874	×		
31	—	—				63	—	—			
32	440	5.657	×	×	×	64	800	8.00	×	×	×

Table A4.2 Occurrence of reflections for the c.p.h. crystal structure

Common allowed reflections			Common forbidden reflections (sometimes these occur by double diffraction, depending upon ***B***)
$10\bar{1}0$	$3\bar{2}\bar{1}1$	$20\bar{2}3$	0001
$1\bar{2}10$	$2\bar{2}01$	$\bar{2}4\bar{2}3$	0003
$20\bar{2}0$	$\bar{2}112$	$10\bar{1}4$	$1\bar{2}11$
$12\bar{3}0$	$01\bar{1}2$	$\bar{2}114$	$1\bar{2}13$
$30\bar{3}0$	0002	$10\bar{1}5$	$\bar{3}033$
$13\bar{4}0$	$11\bar{2}2$	$\bar{1}2\bar{1}6$	$1\bar{2}15$
$5\bar{1}40$	$21\bar{3}2$		
$01\bar{1}1$	$10\bar{1}3$		

APPENDIX 6. INTERPLANAR ANGLES AND SPACINGS OF SELECTED MATERIALS

Tables A6.1–A6.6 list interplanar spacings and angles for various materials. Further information may be obtained in Andrews *et al.* (1971). For interplanar angles, the specific combinations of indices for the two planes may be obtained with the equation for cos ϕ, table A1.1, for the cubic crystal structure.

Appendix 6: References

Andrews, K. W., Dyson, D. J., and Keown, S. R. (1971). *Interpretation of Electron Diffraction Patterns*, Hilger-Watts, London.

Barrett, C. S. (1971). *Structure of Metals*, McGraw-Hill, New York.

Table A6.1 Angles between crystallographic planes (and between crystallographic directions) in crystals of the cubic system [After Barrett (1971)]

{*HKL*}	{*hkl*}	Values of angles between *HKL* and *hkl* planes (or directions)						
100	100	0°	90°					
	110	45°	90°					
	111	54°44′						
	210	26°34′	63°26′	90°				
	211	35°16′	65°54′					
	221	48°11′	70°32′					
	310	18°26′	71°34′	90°				
	311	25°14′	72°27′					
	320	33°41′	56°19′	90°				
	321	36°42′	57°41′	74°30′				
110	110	0°	60°	90°				
	111	35°16′	90°					
	210	18°26′	50°46′	71°34′				
	211	30°	54°44′	73°13′	90°			
	221	19°28′	45°	76°22′	90°			
	310	26°34′	47°52′	63°26′	77°5′			
	311	31°29′	64°46′	90°				
	320	11°19′	53°58′	66°54′	78°41′			
	321	19°6′	40°54′	55°28′	67°48′	79°6′		
111	111	0°	70°32′					
	210	39°14′	75°2′					
	211	19°28′	61°52′	90°				
	221	15°48′	54°44′	78°54′				
	310	43°6′	68°35′					
	311	29°30′	58°31′	79°59′				
	320	36°49′	80°47′					
	321	22°12′	51°53′	72°1′	90°			
210	210	0°	36°52′	53°8′	66°25′	78°28′	90°	
	211	24°6′	43°5′	56°47′	79°29′	90°		
	221	26°34′	41°49′	53°24′	63°26′	72°39′	90°	
	310	8°8′	31°57′	45°	64°54′	73°34′	81°52′	
	311	19°17′	47°36′	66°8′	82°15′			
	320	7°7′	29°45′	41°55′	60°15′	68°9′	75°38′	82°53′
	321	17°1′	33°13′	53°18′	61°26′	68°59′	83°8′	90°
211	211	0°	33°33′	48°11′	60°	70°32′	80°24′	
	221	17°43′	35°16′	47°7′	65°54′	74°12′	82°12′	
	310	25°21′	49°48′	58°55′	75°2′	82°35′		
	311	10°1′	42°24′	60°30′	75°45′	90°		
	320	25°4′	37°37′	55°33′	63°5′	83°30′		
	321	10°54′	29°12′	40°12′	49°6′	56°56′		
		70°54′	77°24′	83°44′	90°			
221	221	0°	27°16′	38°57′	63°37′	83°37′	90°	
	310	32°31′	42°27′	58°12′	65°4′	83°57′		
	311	25°14′	45°17′	59°50′	72°27′	84°14′		
	320	22°24′	42°18′	49°40′	68°18′	79°21′	84°42′	
	321	11°29′	27°1′	36°42′	57°41′	63°33′	74°30′	
		79°44′	84°53′					

Table A6.1 (*continued*)

{*HKL*}	{*hkl*}	Values of angles between *HKL* and *hkl* planes (or directions)						
310	310	0°	25°51′	36°52′	53°8′	72°33′	84°16′	90°
	311	17°33′	40°17′	55°6′	67°35′	79°1′	90°	
	320	15°15′	37°52′	52°8′	58°15′	74°45′	79°54′	
	321	21°37′	32°19′	40°29′	47°28′	53°44′	59°32′	
		65°		85°9′	90°			
311	311	0°	35°6′	50°29′	62°58′	84°47′		
	320	23°6′	41°11′	54°10′	65°17′	75°28′	85°12′	
	321	14°46′	36°19′	49°52′	61°5′	71°12′	80°44	
320	320	0°	22°37′	46°11′	62°31′	67°23′	72°5′	90°
	321	15°30′	27°11′	35°23′	48°9′	53°37′	58°45′	68°15′
		72°45′	77°9′	85°45′	90°			
321	321	0°	21°47′	31°	38°13′	44°25′	50°	60°
		64°37′	69°4′	73°24′	81°47′	85°54′		

Table A6.2 Interplanar spacings of selected materials with f.c.c. crystal structure

	γFe	Cu	Pt	Al	Au	Ag	Pb	Ni	Co (β)
Lattice parameter a_0 (Å)	3.5852	3.6150	3.9231	4.0496	4.0780	4.0862	4.9505	3.5238	3.5520
hkl	*d* spacing (Å)	*d* spacing (Å)	*d* spacing (Å)	*d* spacing (Å)	*d* spacing (Å)	*d* spacing (Å)	*d* spacing (Å)	*d* spacing (Å)	*d* spacing (Å)
111	2.070	2.087	2.265	2.338	2.355	2.359	2.858	2.0345	2.0508
002	1.793	1.808	1.962	2.025	2.039	2.044	2.475	1.7619	1.7760
022	1.268	1.278	1.387	1.432	1.442	1.445	1.750	1.2460	1.2560
113	1.081	1.090	1.183	1.221	1.230	1.231	1.493	1.0623	1.0708
222	1.035	1.044	1.133	1.169	1.177	1.180	1.429	1.0172	1.0254
004	0.896	0.904	0.981	1.012	1.020	1.022	1.238	0.8810	0.8880
133	0.823	0.829	0.900	0.929	0.936	0.938	1.136	0.8084	0.8149
024	0.802	0.808	0.877	0.906	0.912	0.914	1.107	0.7880	0.7943
224	0.732	0.738	0.801	0.827	0.832	0.834	1.011	0.7193	0.7250
333 } 115 }	0.690	0.696	0.755	0.779	0.785	0.786	0.953	0.6782	0.6836
044	0.634	0.639	0.694	0.716	0.721	0.722	0.875	0.6229	0.6279
135	0.606	0.610	0.663	0.685	0.689	0.691	0.837	0.5956	0.6004
006 } 244 }	0.598	0.603	0.654	0.675	0.680	0.681	0.825	0.5873	0.5920
026	0.567	0.572	0.620	0.640	0.645	0.646	0.783	0.5571	0.5616
335	0.547	0.551	0.598	0.618	0.622	0.623	0.755	0.5374	0.5417
226	0.541	0.545	0.591	0.611	0.615	0.616	0.746	0.5313	0.5355
444	0.518	0.522	0.566	0.585	0.589	0.590	0.715	0.5086	0.5127

Table A6.3 Interplanar spacings of selected materials with b.c.c. crystal structure

	αFe	Cr	Mo	W	Nb	Ta	V
Lattice parameter a_0 (Å)	2.8661	2.8850	3.1463	3.1652	3.3007	3.3058	3.0390
hkl	*d* spacing (Å)	*d* spacing (Å)	*d* spacing (Å)	*d* spacing (Å)	*d* spacing (Å)	*d* spacing (Å)	*d* spacing (Å)
011	2.027	2.040	2.225	2.238	2.334	2.338	2.149
002	1.433	1.443	1.573	1.583	1.650	1.653	1.5195
112	1.170	1.178	1.285	1.292	1.348	1.350	1.2409
022	1.013	1.020	1.113	1.119	1.167	1.169	1.0746
013	0.906	0.912	0.995	1.001	1.044	1.045	0.9611
222	0.828	0.833	0.908	0.914	0.953	0.954	0.8773
123	0.766	0.771	0.841	0.846	0.882	0.884	0.8121
004	0.717	0.721	0.787	0.791	0.825	0.826	0.7598
114 } 033 }	0.676	0.680	0.742	0.746	0.778	0.779	0.7162
024	0.641	0.645	0.704	0.708	0.738	0.739	0.6796
233	0.611	0.615	0.671	0.675	0.704	0.705	0.6480
224	0.585	0.589	0.642	0.646	0.674	0.675	0.6203
015 } 134 }	0.562	0.566	0.617	0.621	0.647	0.648	0.5960

Table A6.4 Interplanar spacings of selected materials with diamond cubic crystal structure

Lattice parameter (Å)	Si 5.4282	Ge 5.6580
(*hkl*)	*d* spacing (Å)	*d* spacing (Å)
111	3.1340	3.2667
220	1.9194	2.0007
311	1.6365	1.7058
400	1.3571	1.4145
331	1.2453	1.2980
422	1.1080	1.1549
511	1.0447	1.0889
440	0.9596	1.0002
531	0.9175	0.9564
620	0.8582	0.8945
533	0.8278	0.8629
444	0.7835	0.8167
711	0.7601	0.7923

Table A6.6 Interplanar spacings of graphite, $a = 2.461$ Å, $c = 6.708$ Å

hkil	*d* spacing (Å)
0002	3.354
$10\bar{1}0$	2.131
$10\bar{1}1$	2.031
$10\bar{1}2$	1.799
0004	1.677
$10\bar{1}3$	1.543
$11\bar{2}0$	1.231
$11\bar{2}2$	1.155
0006	1.118

Table A6.5 Interplanar spacings of selected materials with close-packed hexagonal crystal structure

		Be	Zn	Ti	Mg	Zr	Gd	Co	Cd	Re
Lattice parameters	a_0 (Å)	2.285	2.664	2.950	3.209	3.231	3.636	2.507	2.979	2.761
	c_0 (Å)	3.584	4.046	4.683	5.210	5.147	5.782	4.069	5.617	4.458
	c_0/a_0	1.568	2.856	1.587	1.593	1.593	1.590			
hkil		*d* spacing (Å)	*d* spacing (Å)	*d* spacing (Å)	*d* spacing (Å)	*d* spacing (Å)	*d* spacing (Å)	*d* spacing (Å)	*d* spacing (Å)	*d* spacing (Å)
0001		3.584	4.947	4.683	5.210	5.148	5.783	4.068	5.617	4.458
$01\bar{1}0$		1.979	2.308	2.555	2.779	2.798	3.149	2.170	2.580	2.390
0002		1.792	2.473	2.342	2.605	2.574	2.891	2.035	2.808	2.229
$01\bar{1}1$		1.733	2.092	2.243	2.452	2.439	2.765	1.915	2.344	2.107
$01\bar{1}2$		1.329	1.687	1.726	1.901	1.894	2.130	1.484	1.900	1.630
0003		1.195	1.649	1.561	1.737	1.716	1.928	1.356	1.872	1.486
$11\bar{2}0$		1.143	1.332	1.475	1.605	1.616	1.818	1.253	1.489	1.380
$11\bar{2}1$		1.089	1.287	1.407	1.534	1.541	1.734	1.198	1.440	1.318
$01\bar{1}3$		1.023	1.342	1.332	1.473	1.463	1.644	1.150	1.515	1.262
$02\bar{2}0$		0.990	1.154	1.278	1.390	1.399	1.574	1.085	1.290	1.195
$11\bar{2}2$		0.964	1.173	1.248	1.366	1.368	1.539	1.067	1.316	1.173
$02\bar{2}1$		0.954	1.124	1.233	1.343	1.350	1.519	1.048	1.257	1.154
0004		0.896	1.237	1.171	1.303	1.287	1.446	1.017	1.404	1.114
$02\bar{2}2$		0.866	1.046	1.122	1.226	1.229	1.383	0.957	1.172	1.053
$11\bar{2}3$		0.826	1.036	1.072	1.179	1.176	1.323	0.920	1.166	1.011
$01\bar{1}4$		0.816	1.090	1.064	1.180	1.169	1.314	0.921	1.233	1.010
$02\bar{2}3$		0.762	0.945	0.989	1.085	1.084	1.219	0.847	1.062	0.931
$12\bar{3}0$		0.748	0.872	0.966	1.051	1.058	1.190	0.820	0.975	0.903
$12\bar{3}1$		0.732	0.859	0.946	1.030	1.036	1.166	0.804	0.961	0.885
0005		0.717	0.989	0.937	1.042	1.030	1.157	0.814	1.123	0.891
$11\bar{2}4$		0.705	0.906	0.917	1.011	1.007	1.132	0.780	1.022	0.867
$12\bar{3}2$		0.690	0.823	0.893	0.974	0.978	1.101	0.761	0.921	0.837
$01\bar{1}5$		0.674	0.909	0.879	0.976	0.966	1.086	0.761	1.030	0.835
$02\bar{2}4$		0.664	0.844	0.863	0.950	0.947	1.065	0.742	1.950	0.815
$03\bar{3}0$		0.660	0.769	0.852	0.927	0.933	1.050	0.723	0.859	0.797
$03\bar{3}1$		0.649	0.760	0.838	0.912	0.918	1.033	0.712	0.850	0.784
$12\bar{3}3$		0.634	0.771	0.821	0.899	0.900	1.013	0.701	0.865	0.772
$03\bar{3}2$		0.619	0.735	0.800	0.873	0.877	0.987	0.682	0.822	0.750
$11\bar{2}5$		0.607	0.794	0.791	0.874	0.868	0.976	0.682	0.897	0.749
0006		0.597	0.825	0.781	0.868	0.858	0.964	0.678	0.936	0.743

APPENDIX 9. SUPERIMPOSED STEREOGRAMS FOR VARIOUS COMMON ORIENTATION RELATIONSHIPS

This appendix contains superimposed stereograms (figures A9.1–A9.6) for two phases with well-defined orientation relationships. These stereograms are particularly useful in interpreting diffraction patterns containing reflections from both precipitate and matrix.

For any given ***B*** in one phase the SADP should contain both its own and second-phase spots. Their positions may be found by constructing the great circle with pole, ***B***, see appendix 1. This contains all reflecting planes, that is those whose normals are perpendicular to ***B***, see figure 2.20(e) and section 2.7.2. An example is shown in figure A9.7 for the Nishyama–Wassermann relationship. ***B*** is $[\bar{1}10]$ in the b.c.c. phase and $[\bar{1}11]$ in the f.c.c. phase. The great circle AB with this pole is shown in

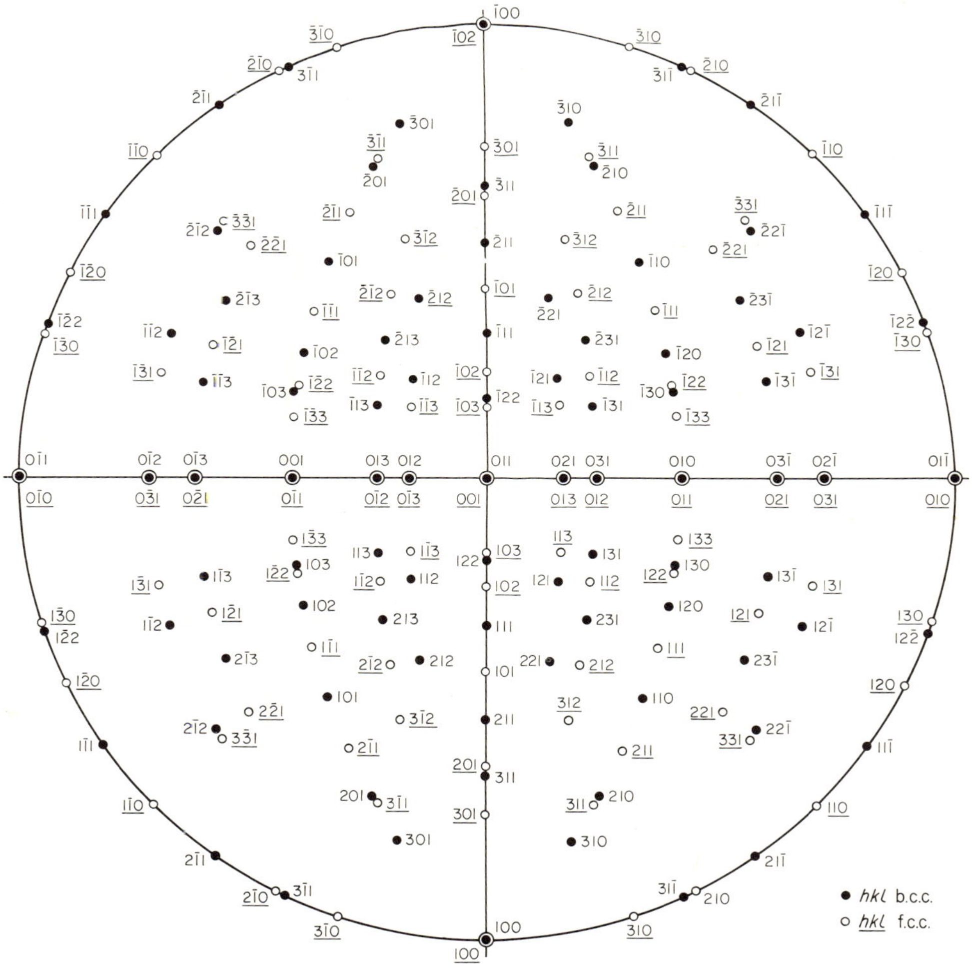

Figure A9.1 Stereographic projection representing the Bain relationship between body- and face-centered cubic materials [After Andrews *et al.* (1971)]
(100) b.c.c. // (100) f.c.c.
$(01\bar{1})$ b.c.c. // (010) f.c.c.
(011) b.c.c. // (001) f.c.c.

figure A9.2 and the superimposed diffraction patterns in figure A9.7. All spots lie in this great circle, distance from the centre spot $\propto 1/d_{hkl}$, and with the correct angular relationships.

It is frequently convenient to prepare standard patterns showing these orientation relationships, for simple ***B*** in the matrix, for example ⟨001⟩, ⟨111⟩, ⟨110⟩, ⟨112⟩, so that these ***B*** may be obtained in a given area by tilting the specimen, thus enabling the orientation relationship to be confirmed, see section 2.11.

Appendix 9: Reference

Andrews, K. W., Dyson, D. J., and Keown, S. R. (1971). *Interpretation of Electron Diffraction Patterns*, Hilger, London.

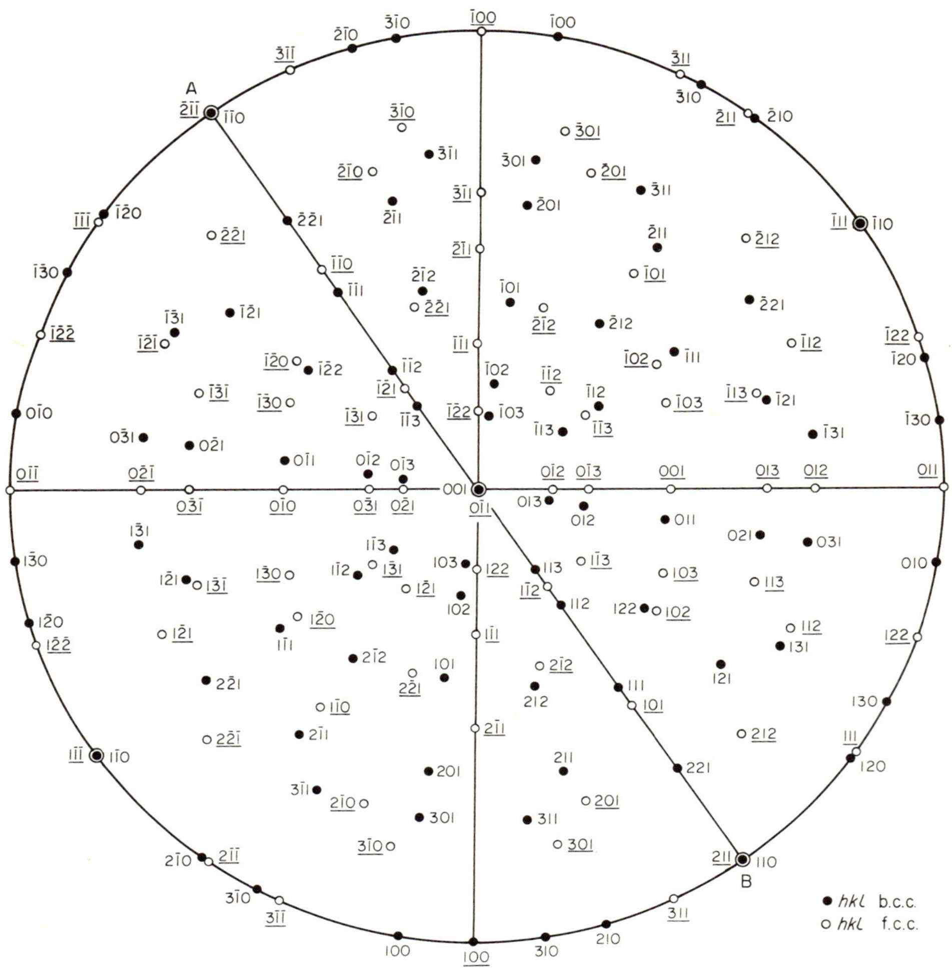

Figure A9.2 Stereographic projection representing the Nishyama–Wassermann orientation relationship between body- and face-centred cubic materials [After Andrews *et al.* (1971)]
(001) b.c.c. // (0$\bar{1}$1) f.c.c.
($\bar{1}$10) b.c.c. // ($\bar{1}$11) f.c.c.
(110) b.c.c. // (211) f.c.c.

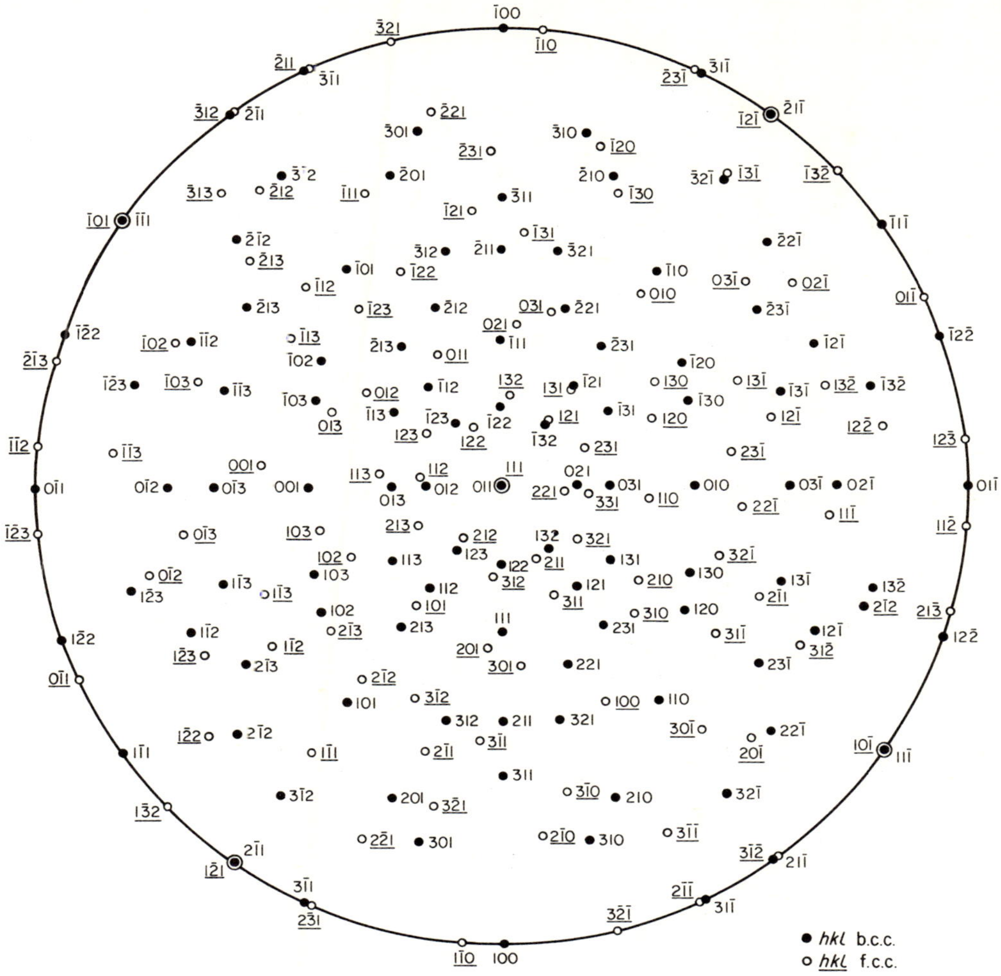

Figure A9.3 Stereographic projection representing the Kurdjumov–Sachs orientation relationship between body- and face-centred cubic materials [After Andrews *et al.* (1971)]

(011) b.c.c. // (111) f.c.c.

(11$\bar{1}$) b.c.c. // (10$\bar{1}$) f.c.c.

($\bar{2}$1$\bar{1}$) b.c.c. // ($\bar{1}$2$\bar{1}$) f.c.c.

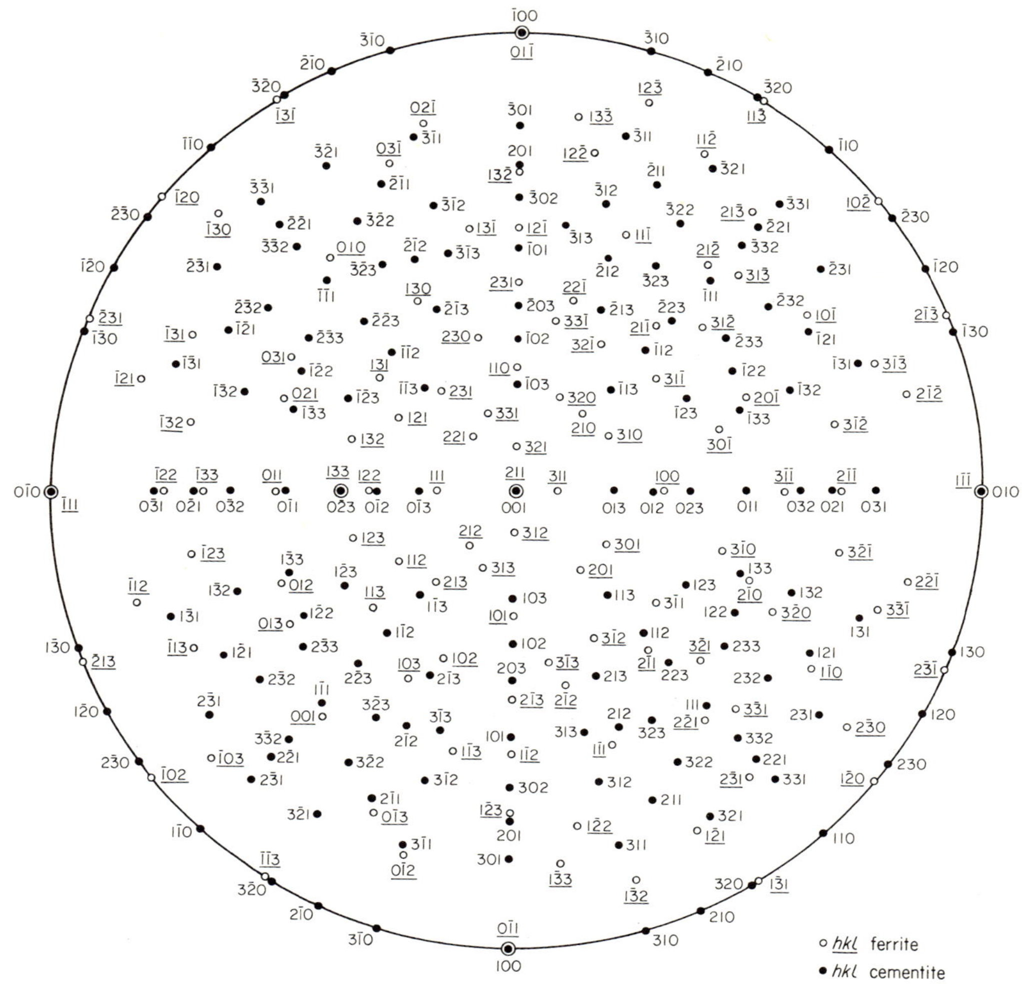

Figure A9.4 Stereographic projection representing the Bagaryatskii orientation relationship between cementite and α-iron [After Andrews *et al.* (1971)]
(100) cementite // (0$\bar{1}$1) α-iron
(010) cementite // (1$\bar{1}\bar{1}$) α-iron
(001) cementite // (211) α-iron

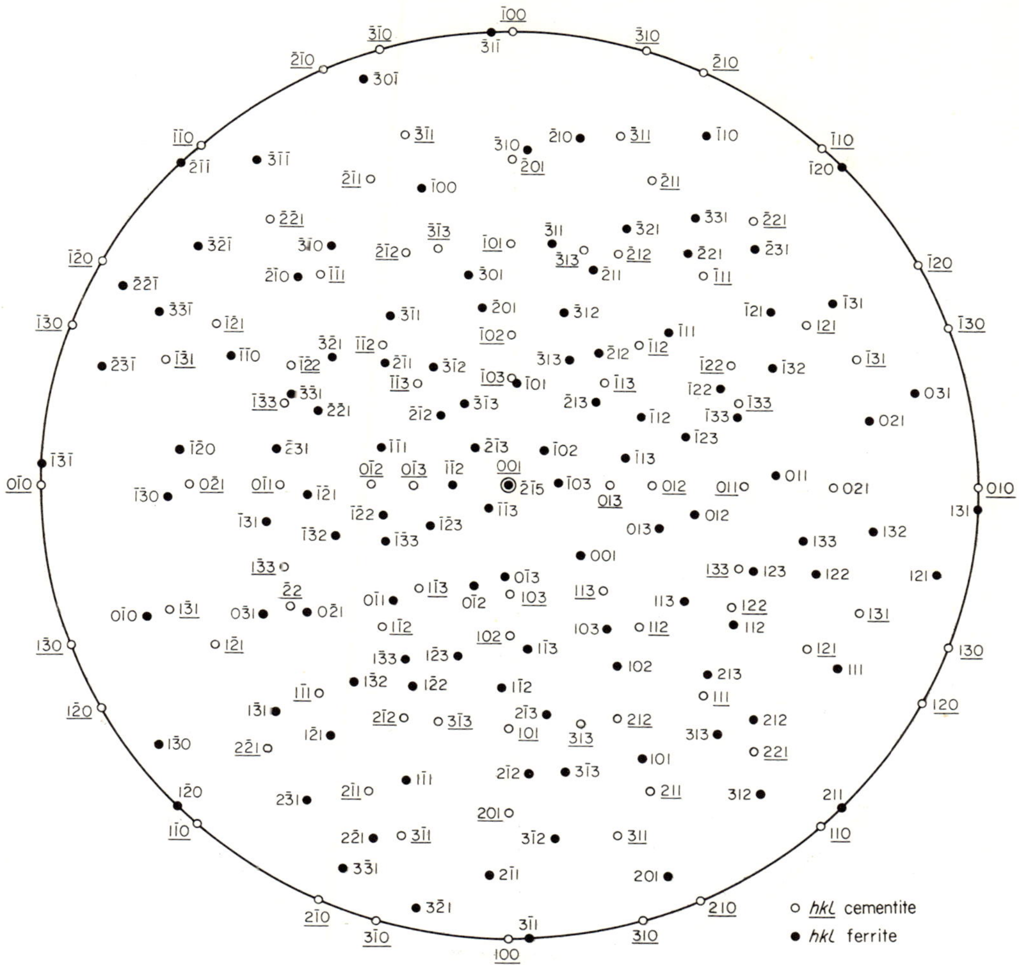

Figure A9.5 Stereographic projection representing the Pitsch orientation relationship between cementite and α-iron. [After Andrews *et al.* (1971).]
(100) cementite // ($3\bar{1}1$) α-iron
(010) cementite // (131) α-iron
(001) cementite // ($\overline{2}\overline{1}5$) α-iron

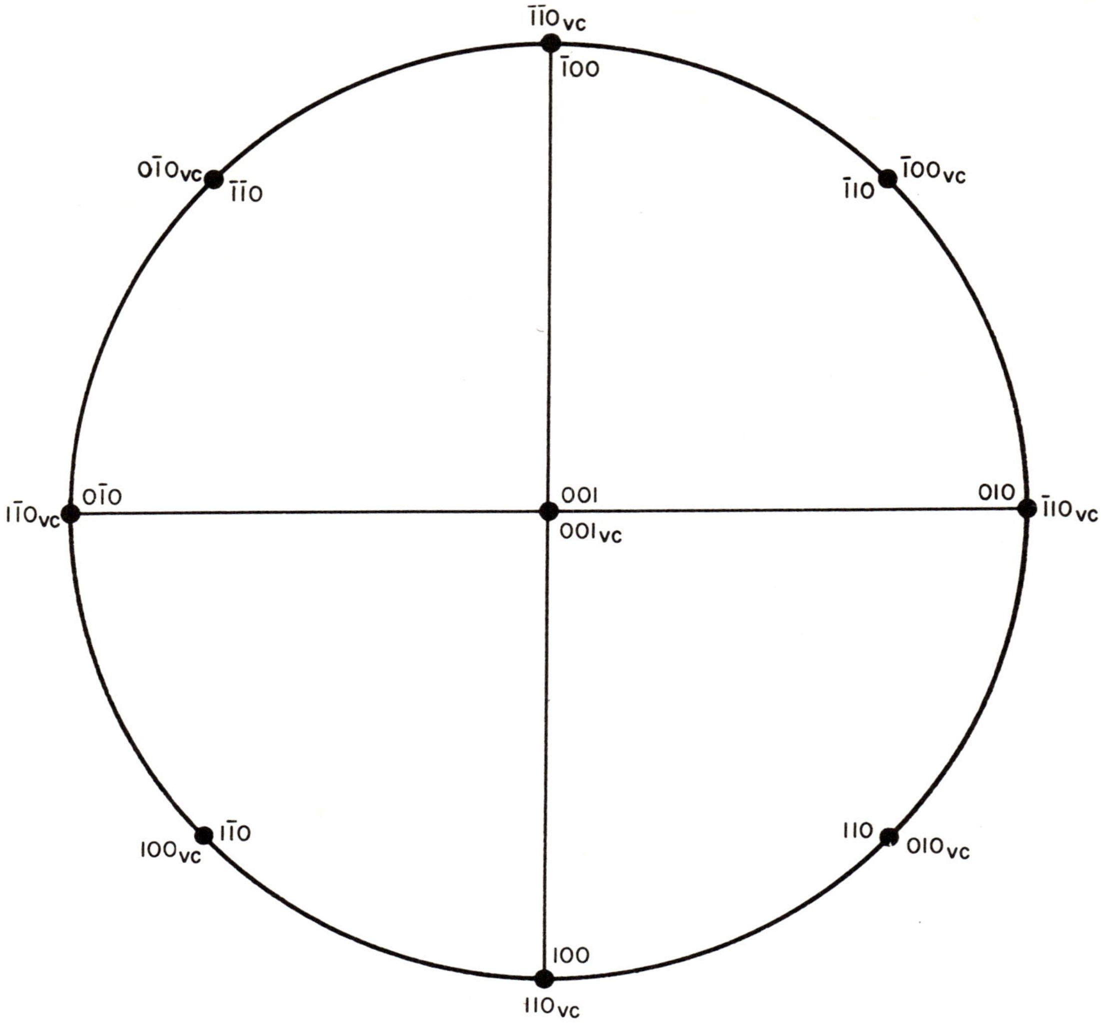

Figure A9.6 Stereographic projection representing the Baker–Nutting orientation relationship between vanadium carbide and α-iron
(001) VC // (001) α-iron
($\bar{1}$10) VC // (010) α-iron
(110) VC // (100) α-iron

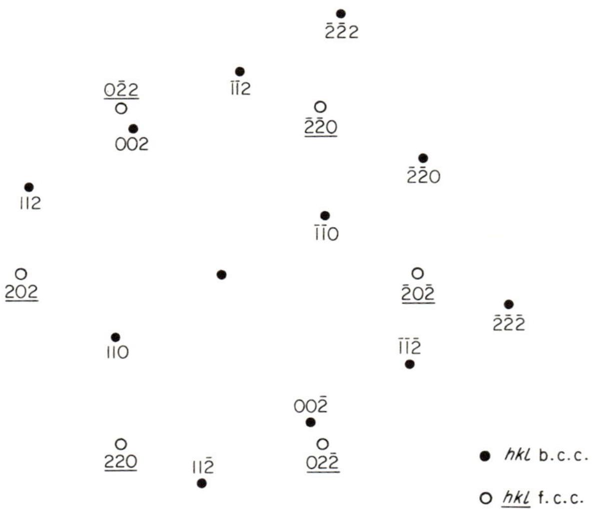

Figure A9.7 Superimposed spot diffraction patterns in which [$\bar{1}$11] f.c.c. // [$\bar{1}$10] b.c.c. = ***B*** for the Nishyama–Wassermann orientation relationship depicted in figure A9.2

APPENDIX 14. ILLUSTRATION OF THE INHERENT AMBIGUITY IN THE INTERPRETATION OF SELECTED AREA ELECTRON DIFFRACTION PATTERNS OF CEMENTITE

The assignment of indices to reflections on a selected area electron diffraction pattern is based on the correlation of interplanar spacing and interangular measurements. Owing to the large unit cell of the orthorhombic cementite structure it is often very difficult to assign the correct indices to reflections in the diffraction pattern since the differences in interplanar spacings of higher-order reflections are very small. It is usually possible to index reflecting zones of low order correctly but even in such cases some difficulty may be encountered.

In order to illustrate this difficulty the $[301]_c$* and $[102]_c$ zones will be considered. These two zones are drawn out in figure A14.1 on the scale of a typical photographic plate.

In table A14.1 the expected values of angles from $(010)_c$ and the distance from the centre spot for each reflection are shown for both the $[301]_c$ and $[102]_c$ zones. $\Delta\theta$ is the angular difference between corresponding reflections. For example, the angular difference between $(\bar{1}33)_c$ and $(\bar{2}31)_c$ is 1.77°. The column, spacing on plate, refers to the actual spacing on the plate (R) between the centre spot and the appropriate reflection. The interplanar spacing (d) is determined from the relation $d = \lambda L/R$ (where λL is the camera constant). The spacing on the plate is considered in this example in order to illustrate the experimental inaccuracy. ΔR is the difference in spacing between corresponding reflections in the two zones. For example, the difference in spacing on the plate between a $(\bar{1}33)_c$ reflection belonging to the $[301]_c$ zone and a $(\bar{2}31)_c$ reflection belonging to the $[102]_c$ zone is 0.5 mm.

Table A14.1

	Reflection		Angle from $(010)_c$		$\Delta\theta$ (deg)	Spacing on plate R (mm)		ΔR(mm)
	$[301]_c$	$[102]_c$	$[301]_c$	$[102]_c$		$[301]_c$	$[102]_c$	
a	010	010	0	0	0	5.07	5.07	0
b	$\bar{1}33$	$\bar{2}31$	40.12	38.35	1.77	19.90	19.40	0.50
c	$\bar{1}23$	$\bar{2}21$	51.65	49.88	1.77	16.35	15.72	0.63
d	$\bar{1}13$	$\bar{2}11$	68.42	67.15	1.27	13.8	13.08	0.72
e	$\bar{1}03$	$\bar{2}01$	90.0	90.0	0	12.85	12.02	0.83

From the tabulated values of table A14.1, it is clear that the maximum angular difference between corresponding reflections is less than 2°, which is very difficult to determine on a diffraction pattern. The maximum difference in distance from the centre spot is 0.83 mm. Bearing in mind that the typical diameter of a cementite reflection is 0.5 mm, it is clear that it is very difficult to distinguish between these two zones. A distinction can only be made if the camera constant is known to a high degree of accuracy, which is not normally the case. The $[301]_c$ zone is, however, 45° away from $[102]_c$ and a method has to be found to enable an unambiguous indexing of reflections. Three different approaches can be taken.

(i) The camera constant can be determined in the case of pearlite diffraction patterns by calculation from a known ferrite reflection in the pattern.

(ii) The foil can be tilted in the microscope so that

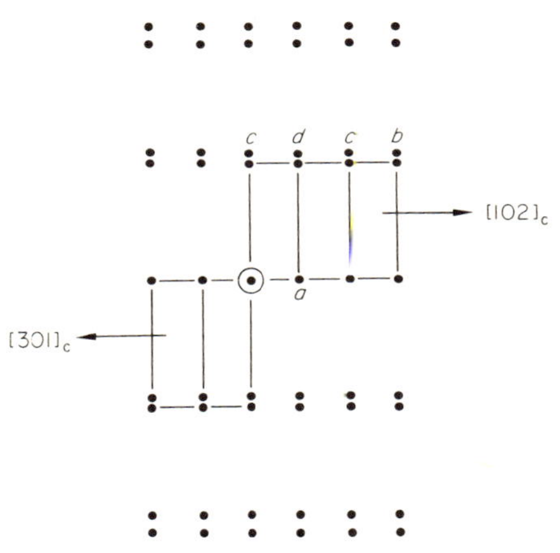

Figure A14.1 Diagram of $[301]_c$ and $[102]_c$ reflecting cementite zones. The possible indices that can be assigned to *a*, *b*, etc., are given in table A14.1

* *Note.* $[uvw]_c$ refers to the zone axis as a crystal direction in cementite and $(hkl)_c$ refers to reciprocal lattice vectors: $hu + kv + lw = 0$.

the Kikuchi centre was in the centre of the pattern. (The actual Kikuchi centre could normally not be seen in pearlite diffraction patterns but a tilting condition was found such that reflections were symmetrical about the centre and of even intensity.) The presence of higher-order reflections belonging to the same zone greatly facilitated in the accurate measurement of the distance of a reflection from the centre spot.

(iii) A second diffraction pattern can be taken from the same area at a different tilting condition. When this new zone was identified and the angle of tilt known, one of the previous possibilities could be ruled out.